T5-AFF-739

Interdisciplinary Food Safety Research

CRC Series in CONTEMPORARY FOOD SCIENCE

Fergus M. Clydesdale, Series Editor
University of Massachusetts, Amherst

Published Titles:

America's Foods Health Messages and Claims: Scientific, Regulatory, and Legal Issues
James E. Tillotson

New Food Product Development: From Concept to Marketplace
Gordon W. Fuller

Food Properties Handbook
Shafiur Rahman

Aseptic Processing and Packaging of Foods: Food Industry Perspectives
Jarius David, V. R. Carlson, and Ralph Graves

The Food Chemistry Laboratory: A Manual for Experimental Foods, Dietetics, and Food Scientists
Connie Weaver

Handbook of Food Spoilage Yeasts
Tibor Deak and Larry R. Beauchat

Food Emulsions: Principles, Practice, and Techniques
David Julian McClements

Getting the Most Out of Your Consultant: A Guide to Selection Through Implementation
Gordon W. Fuller

Antioxidant Status, Diet, Nutrition, and Health
Andreas M. Papas

Food Shelf Life Stability
N.A. Michael Eskin and David S. Robinson

Bread Staling
Pavinee Chinachoti and Yael Vodovotz

Interdisciplinary Food Safety Research
Neal M. Hooker and Elsa A. Murano

Interdisciplinary Food Safety Research

Edited by
Neal H. Hooker
Elsa A. Murano

CRC Press
Boca Raton London New York Washington, D.C.

Cover art based on material from Chapters 5 and 8.

Library of Congress Cataloging-in-Publication Data

Interdisciplinary food safety research / edited by Neal H. Hooker and Elsa A. Murano.
p. cm. -- (Contemporary food science series)
Includes bibliographical references and index.
ISBN 0-8493-2217-0 (alk. paper)
1. Food adulteration and inspection. 2. Food industry and trade--Safety measures.
I. Murano, E.A. (Elsa A.) II. Title. III. Series

TX531 .H817 2001
363.19'26--dc21 00-046757

Direct all inquiries to CRC Press LLC, 2000 N.W. Corporate Blvd., Boca Raton, Florida 33431.

International Standard Book Number 0-8493-2217-0
Library of Congress Card Number 00-046757
Printed in the United States of America 1 2 3 4 5 6 7 8 9 0
Printed on acid-free paper

Foreword

It is a pleasure to provide a foreword to this important book. The authors are both known and respected in the field of food safety, and this text will materially add to the body of knowledge about how, and under what conditions, food may be considered safe. I congratulate them on this impeccably researched and well-written work.

Modern food safety is a comparatively recent phenomenon. Of course, there could not have been a rational approach to the production of safe food before the advent of the germ theory of disease in the latter half of the nineteenth century, but knowing that the microscopic creatures observed for centuries had the capacity to cause disease through the vehicle of food did not yield a ready answer to the problem. What was needed was a scheme for preventing foodborne disease based on the startling new findings. Sadly, it would take over 100 years to consolidate the science into a prevention-oriented infrastructure.

Looking back with the advantage of subsequent history, the early attempts to bring order to this area of preventive medicine were woeful and even childlike in design. The organoleptic method of safety assurance was predicated on the false precept that foodborne disease could be prevented by physical examination of food.

In addition to missing a great deal of potentially dangerous contamination, vast quantities of good food were summarily rejected. Strangely, the essential instrument of food safety evaluation, the microscope, was rarely employed in early food control programs.

A few epochal events transformed these erratic neo-nascent safety programs into a better organized, more effective system. Among these was the conquest of botulism by the canning industry in the 1920s. This involved scientific investigation and the development of a control process that remains one of the true jewels of medical history. Previous work had identified the spore-forming bacterium *Clostridium botulinum* as the causative agent. It remained for the genius of Dr. K.F. Meyer of the University of California-San Francisco to develop the retort process that provided reasonable assurance of no harm, thereby saving the canning industry. Far more importantly, this event provided the methodology that has served as the model for developing specific foodborne pathogen control systems. Earlier, Pasteur had introduced the concept of a kill step with his pioneering work on the pasteurization of beer, and chlorination of water provided the first truly useful model for chemical decontamination.

Ultimately, the multitude of individual pathogen-oriented food control programs were unified in the Hazard Analysis Critical Control Point Program (HACCP). Pillsbury's Dr. Howard Bauman developed this systems approach of food safety first for the space program, but later it became the acknowledged world standard for safety assurance and food inspection programs.

What is next for food safety? Some believe there will be no next. HACCP, they maintain, is as good as it will ever get. In this they are wrong. The legatees of Pasteur and Meyer, such as the authors of this book, will not long be satisfied with HACCP, because the lesson of history is that science and medicine always demand improvement. Moreover, it would be both unthinkable and unethical not to strive for something better.

I personally think that the next generation of food safety is to be found in the food safety objective (FSO) concept. This approach focuses an overall program on human disease-reduction goals rather than the more indirect HACCP system, which presumes that the reduction of foodborne pathogens in food at some finite point in the food chain will inevitably lead to a diminution of human disease.

Notwithstanding all the above, whatever the future holds for the practice of food safety, the science leading the way to better, more effective systems reposes in this book. The obviously painstaking compilation of this impressive array of relevant information can only lead to the means whereby earnest workers in the field can think through the manifold considerations that must be assimilated in the all-important calling that is the food safety of today . . . and tomorrow. K.F. Meyer would have liked this book, as will you.

Lester M. Crawford, D.V.M., Ph.D.
Director, Center for Food and Nutrition Policy
Georgetown University
Washington, D.C.

Preface

Food safety concerns have received increasing attention from agribusinesses, government agencies, consumers, and researchers over the last few years. However, much of the previous research has focused on single disciplinary aspects of what we now understand to be a complex and holistic set of problems. The intent of the editors in preparing *Interdisciplinary Food Safety Research* was to address this shortcoming. We became aware of a number of research groups around the world who were constructing novel interdisciplinary teams to consider complex food safety issues that were beyond the traditional disciplinary arenas. Indeed, our own experiences in developing such a team were at the same time rewarding and somewhat frustrating. We felt that it would have been easier to perform the research if we could follow a set of guidelines of tools, team logistics, and methods to encourage stakeholder interactions prepared by others who had experience with such groups. The lack of such a truly interdisciplinary reference was the motivation for proposing this text. We hope that the lessons of others can suggest certain techniques and methods to combine across the necessary range of disciplinary expertise. Thereby, the reader should be better placed to form his or her own interdisciplinary team.

Along with this desired goal of providing some guidance to other groups, we observed a paradox within the general arena of interdisciplinary food safety research. While we firmly believe that only by combining the research skills of a range of disciplines can many of the more pressing food safety concerns be addressed, it is notoriously difficult to publish such work in disciplinary academic journals. This problem becomes more pronounced with the increasing diversity of team members, either by discipline or background (academic, government, or industry). Therefore, we hope to provide, in this volume, an outlet for a preliminary set of excellent work that we feel deserves such recognition. The research presented in this book spans a range of food safety problems, disciplines, and stages of the food chain. Authors are drawn from a range of institutional backgrounds and nations to highlight the diverse nature of food safety issues and to motivate a holistic evaluation wherever possible. We hope that this volume will encourage further exploration of truly interdisciplinary food safety research. We expect there will be an increasing supply of such excellent interdisciplinary research in the coming years and encourage all to develop their own teams and thereby enjoy the challenge!

The Editors

Neal H. Hooker earned a Ph.D. in resource economics from the University of Massachusetts in 1997. His dissertation considered the impacts of international trade agreements such as the Uruguay Round of GATT and NAFTA on food safety and competitiveness. Dr. Hooker then completed a postdoctoral research associate program at the University of Massachusetts, with the research extending his dissertation studies. He concurrently held a postdoctorate in the Center for Food Safety at Texas A&M University. That position primarily involved performing a cost benefit analysis of pathogen reduction strategies for Australian beef slaughter plants. He was also appointed as an Adjunct Assistant Professor in the Department of Agricultural Economics at Texas A&M. Dr. Hooker is an active member in the American Agricultural Economics Association, the International Food and Agribusiness Marketing Association, and the Food Distribution Research Society.

Dr. Hooker next became an Assistant Professor at the Department of Agricultural and Resource Economics at Colorado State University, where he taught an undergraduate course in agricultural marketing and co-taught a graduate agricultural policy class. Subsequently, Dr. Hooker moved to the Department of Agricultural, Environmental, and Development Economics at The Ohio State University, where he is currently an Assistant Professor.

He has published ten journal articles and six book chapters on the economics of food safety and quality, considering aspects such as the role of HACCP as an international trade standard, the international marketing of food safety, and comparisons of voluntary and mandatory quality management systems. He is currently co-editing a special issue of a journal on "Private Sector Management of Food Safety." Current research projects include evaluations of quality management systems in the fresh cut (produce) industry; work in the areas of biotechnology, organic production, and the application of novel techniques in agri-food firms; tracking the domestic and international impacts of HACCP programs on meat and poultry firms; policy and trade effects of international trade agreements; and determining the financial implications of food recalls.

Elsa A. Murano earned an M.S. in anaerobic bacteriology in 1987 and a Ph.D. in food science in 1990, both from Virginia Tech. She served as Assistant Professor in charge of research programs at the Iowa State University Linear Accelerator, where she led several projects on the use of food irradiation. Dr. Murano has an active research program on the safety of foods, with funded projects ranging from the study of intervention strategies for the decontamination of fresh meats, fruits, and vegetables, to the prevalence and virulence of specific foodborne pathogens such as *Listeria,* verotoxigenic *Escherichia coli,* and *Aerobacter.* Dr. Murano has been an invited speaker on food safety and irradiation on numerous occasions, both nationally and internationally, and has conducted dozens of HACCP workshops in the U.S. and Latin America as a lead instructor, certified by the International HACCP Alliance. She serves on the editorial boards of several food safety journals and is chair of the Operations Committee of the National Alliance for Food Safety. Dr. Murano is the editor and co-author of *Food Irradiation: A Sourcebook,* ISU Press. She is also co-author of two chapters in the *Compendium of Methods for the Microbiological Examination of Foods.* Dr. Murano currently serves as Associate Professor of Food Science in the Department of Animal Science, and is the director of the Center for Food Safety at Texas A&M University. She is also the current holder of the Sadie Hatfield Endowed Professorship in Agriculture at Texas A&M.

Contributors

Gary R. Acuff
Department of Animal Science and Center for Food Safety
Texas A&M University
College Station, Texas

Hugh Bigsby
Commerce Division
Lincoln University
Canterbury, New Zealand

Christine Bruhn
Center for Consumer Research
University of California
Davis, California

Jean C. Buzby
Economic Research Service
U.S. Department of Agriculture
Washington, D.C.

Stephen Carroll
Supply Chain Consultants
North Sydney, Australia

Suzi Barletto Cavalli
Department of Food Planning and Nutrition
State University of Campinas
São Paulo, Brazil

Andrew Fearne
Food Management Group
Wye College, University of London,
Wye, Ashford Kent, U.K.

John Fox
Department of Agricultural Economics
Kansas State University
Manhattan, Kansas

Paul D. Frenzen
Economic Research Service
U.S. Department of Agriculture
Washington, D.C.

Chris Griffith
Food Research and Consultancy Unit
School of Applied Sciences
University of Wales Institute
Cardiff, U.K.

Robert O. Herrmann
Department of Agricultural Economics and Rural Sociology
The Pennsylvania State University
State College, Pennsylvania

Jill E. Hobbs
Department of Agricultural Economics
University of Saskatchewan
Saskatoon, Canada

Neal H. Hooker
Department of Agricultural, Environmental, and Development Economics
The Ohio State University
Columbus, Ohio

David Lloyd
Food Industry Centre
School of Applied Sciences
University of Wales Institute
Cardiff, U.K.

Scott Malcolm
Food and Resource Economics
University of Delaware
Newark, Deleware

Vartguess Markarian
Dames & Moore
North Sydney, Australia

Adriana Zenlotti Mercadante
Department of Food Science
State University of Campinas
São Paulo, Brazil

Mohammad Modarres
Center for Technology Risk Studies
University of Maryland
College Park, Maryland

Matthew Mortlock
Food Research and Consultancy Unit
School of Applied Sciences
University of Wales Institute
Cardiff, U.K.

Elsa A. Murano
Department of Animal Science and Center for Food Safety
Texas A&M University
College Station, Texas

Flávia Maria Netto
Department of Food Planning and Nutrition
State University of Campinas
São Paulo, Brazil

Clare Narrod
U.S. Department of Agriculture
Office of Risk Assessment and Cost Benefit Analysis
Washington, D.C.

José Luiz Pereira
Department of Food Science
State University of Campinas
São Paulo, Brazil

Adrian Peters
Food Research and Consultancy Unit
School of Applied Sciences
University of Wales Institute
Cardiff, U.K.

Barbara Rasco
Department of Food Science and Human Nutrition
Washington State University
Pullman, Washington

Tanya Roberts
U.S. Department of Agriculture
Economic Research Service
Washington, D.C.

Elisabete Salay
Department of Food Planning and Nutrition
State University of Campinas
São Paulo, Brazil

Stephen Sapp
Department of Sociology
Iowa State University
Ames, Iowa

John Spriggs
Faculty of Science and Agriculture
Charles Stuart University
Wagga Wagga, Australia

Arthur Sterngold
Department of Business Administration
Lycoming College
Williamsport, Pennsylvania

Rex H. Warland
Department of Agricultural Economics and Rural Sociology
The Pennsylvania State University
State College, Pennsylvania

Carolyn Whyte
Ministry of Agriculture and Forestry
MAF Quarantine Service
Auckland, New Zealand

Contents

Chapter 1

An Interdisciplinary Approach to Developing a Probabilistic Risk Analysis Model: Applications to a Beef Slaughterhouse

Tanya Roberts, Clare Narrod, Scott Malcolm, and Mohammad Modarres

Risk assessment has evolved from straightforward identification of hazards to complex models incorporating probability distributions and uncertainty of knowledge. Each discipline brings different methods, models, and data sources to risk assessment. The mix of the disciplines (such as decision science, engineering, economics, food science, and epidemiology) involved in building a risk assessment for foodborne pathogens is likely to influence the design and implementation of the model. By examining models used in engineering and other disciplines, food safety risk assessors can learn about model structure

0-8493-2217-0/01/$0.00+$.50

and variables affecting reliability. Knowledge about the food industry and plant practices is also needed, so a food safety Probabilistic Risk Assessment (PRA) team is wise to include food scientists, veterinarians, and economists along with modelers.

In this chapter, we describe a step-by-step process of building a PRA model for pathogen control. An example of an event tree is developed for a beef slaughterhouse. Each node in the tree indicates the possibility of contamination or decontamination as the animal/carcass moves through the slaughter plant. Several strategies to minimize carcass contamination are possible, and food safety managers need to evaluate these options. We analyze one set of potential risk-cost tradeoffs and construct an efficient frontier of risk reducing strategies.

PRA models can make important contributions to food safety improvement programs, such as Hazard Analysis Critical Control Point (HACCP) systems, by pinpointing where risks can be controlled or eliminated. By combining economics with the PRA model, public and private decision-makers can be better informed about policy options and the likely risk-cost tradeoffs of these policy options.

Introduction

The National Academy of Sciences defines risk assessment as having four components: identifying a hazard, gauging its potency, estimating the likelihood that a group of people have been exposed to the hazard, and characterizing the risk and determining the magnitude of its consequences (NRC, 1983). Risk assessment is, by its very nature, multidisciplinary. It relies on a mixture of scientific, technical, economic, and other information regarding the nature of hazards and how they are produced and can be controlled, as well as information on how individuals, groups, or resources can be exposed to risks, and the magnitude of the impact. The output of a risk assessment can aid policymakers in their decision making concerning alternative control options. Within the rulemaking framework, risk analysis, regulatory impact analysis, and cost-benefit analysis are often closely linked. They need to be closely linked to rationalize the regulatory review process and facilitate policymaking.

Each discipline brings different methods, models, and data sources to risk assessment. The mix of disciplines (such as decision science, engineering, economics, food science, and epidemiology) involved in building a risk assessment model for foodborne pathogens is likely to influence the design and implementation of the model. While a multidisciplinary approach to modeling is necessary, establishing multidisciplinary teams cannot be done without addressing some important concerns. One concern is whether the mix of disciplines is sufficient to capture the complexity of what is being modeled. A second concern is communication and the ability to combine various skills, techniques, models, and lexicons, and to judge the quality of these inputs. A third concern is how to make the risk assessment clear to policy makers so that it can be understood, accepted, and integrated into their decision making. Despite these difficulties, a multidisciplinary approach that incorporates different skill sets enables a risk assessment team to overcome disciplinary limitations and more accurately model the real world.

Risk assessment models vary in their complexity. Some models relating to food safety concerns are

1. Screening risk scenarios to determine if risks from a particular foodborne hazard reach a threshold of concern
2. Ranking relative risks from different foodborne hazards, such as pesticides, parasites, hormone residues, bacteria, and food additives
3. Estimating frequency of occurrence of a particular hazard at a particular location or at all locations from farm to fork
4. Determining risk-significant contributors to develop control mechanisms and define needs for more data
5. Providing the results of the risk assessment to public or private decision makers.

The output of the risk assessment model can be combined with economic information to determine the cost-effectiveness of various risk reduction options in either the public or private sector, provided the variability of effectiveness of the technologies has been incorporated into the risk assessment model. Models that have integrated risk assessment with the costs of pathogen control options include Jensen, Unnevehr, and Gomez (1999), Roberts, Malcolm, and Narrod (1999), and Narrod et al. (1999).

In this chapter we discuss the concepts of PRA for foodborne pathogens and how to develop a PRA model. Then we apply the technique and build a PRA model for the beef slaughterhouse. The model captures how the state and level of microbial contamination changes on a beef carcass at various points in the slaughterhouse. The output of the system is the likely contamination of beef trim destined for grinding into hamburger. An earlier farm-to-table risk assessment for *Escherichia coli* O157:H7 (Cassin et al., 1998) treated the slaughterhouse as a "black box" due to lack of data, and consequently found minimal control opportunities there. Our model explicitly considers the effectiveness of pathogen control technologies at several steps in the slaughter process. The pathogen in our illustrative model is generic *E. coli*[1] because it is a useful indicator of the quality of statistical process control within a slaughter plant (Bisaillon et al., 1997). We illustrate one approach to build the model and test its sensitivity.

We conclude by discussing a technique to integrate the impact of control technologies on the carcass' pathogen load with the cost of the technologies. There are several options to reduce pathogen risk within a slaughterhouse. Some combinations of these options may prove to be more cost effective than others. We demonstrate how PRA and cost/benefit analysis (CBA) can be linked to aid in the policymaking process.

Probabilistic Risk Assessment Methodology

This section examines the basic elements of PRA, explaining when and how the various steps must be performed. PRA models have been used in a variety of applications over the past several decades. Applications have included evaluating "what if" scenarios of nuclear proliferation and atomic war, the probability of nuclear power plant meltdowns or less serious failures, and analysis of the risks in financial portfolios. Only recently have probabilistic techniques been used in a farm-to-table context for foodborne pathogens.

Define Scope of Probabilistic Risk Assessment

A PRA begins with a definition of the scope and objectives of the PRA. The objectives should be clearly stated and achievable with available resources. A typical objective might be to build a probabilistic model to establish distribution of the levels of microbial contamination in slaughterhouse products. The scope of the analysis involves setting boundaries and determining the level of detail necessary to build an accurate model. A key question to ask is what elements of the operation are influential in producing the measurable outcome? Not all functions will be important with respect to achieving the objective. Setting boundaries is equally important so that readers understand what is included and excluded in the model. For example, the output of a beef slaughterhouse usually consists of prime cuts and containers of beef trim for further processing into ground beef products. Some plants produce ground beef, while others

---

[1] Useful models can be developed for either foodborne pathogens or for indicators of pathogens. Pathogens are a more direct link to the probability of human illness, but their numbers may be so limited that costs of monitoring for a particular pathogen may be prohibitively expensive. Indicator organisms, such as total numbers of coliforms or generic *E. coli,* may be easier to locate in a testing program and may be reliable indicators of whether a process in the plant is working as planned. For example, sampling beef carcasses after the hide is removed and testing for indicator organisms can certainly identify gross deficiencies in sanitary hide removal. The test cannot determine whether a particular pathogen, perhaps *E. coli* O157:H7, is present on the carcass.

transport the beef trim to other facilities. The output for our PRA model is the production of beef trim, and any further processing within a beef slaughterhouse is not considered.

An inventory of possible resources for the desired analysis should be developed. Appropriate resources might include computer equipment and software, experts in facility management and operation, experts in engineering and statistical analysis, and experts in animal microbiology. An estimate must be made of when and where these resources are needed. Some of these resources will be members of the core team involved in the PRA on a day-to-day basis, while others will act in a consulting role. Combining these elements, in essence, provides a road map for the project.

A detailed examination of the hazard and the system or process (e.g., specific to the facility, firm, or industry of concern), administrative controls, as well as protective and preventive systems is required. All systems, facilities, processes, and activities expected to play a role in the initiation, propagation, or arrest of a hazard exposure or hazardous condition must be understood in sufficient detail to construct the models necessary to capture all possible exposure pathways.

Steps to identify the important pathways for the hazard of interest include:

1. Identify major preventive or corrective systems (or methods) available to control the hazard.
2. Describe the physical interactions among all major systems, processes, and human operators. These relationships could be summarized in a dependency matrix.
3. Study past major events and use the information to help ensure inclusion of important applicable scenarios.
4. Initiate a dependable recordkeeping system. Consistent documentation is key to ensuring the quality of the PRA.
5. Define the scope of the PRA, list which faults and conditions are included or excluded, and identify the primary modes of concern.

Once the details of the PRA are laid out, the core team should determine the ground rules for the analysis and assign responsibility for each step. This task can be accomplished with the help of outside experts and other stakeholders. Regular communication between participants is vital to ensure smooth and timely information flow and carrying out of responsibilities.

Develop Scenarios

To build the scenarios that could lead to exposure to a hazard, analysts must first identify those "initiating events" (abnormal events) that could, if not correctly controlled, result in hazard exposure. The first step involves identifying sources of hazard and existing barriers around these hazards (or ways their impacts are minimized).[2] The next step involves identifying events that can lead to a direct threat to the integrity of the barriers or control actions. The goal of scenario development is to derive a complete set of scenarios to encompass all potential propagation paths that can lead to loss of confinement of the hazard following occurrence of an initiating event. To describe the cause-and-effect relationship between initiators and event progression, it is necessary to identify those functions that must be maintained to prevent or mitigate hazard exposure. For example, careful hide removal in the slaughterhouse is a preventive measure to reduce the chance of the sterile carcass becoming contaminated with pathogens that might be on the hide.

The sequences of events that define scenarios are frequently displayed by event trees. Event trees order and depict (in an approximately chronological manner) the success or failure of key mitigating actions (e.g., human, equipment, software, or system monitoring actions) that are

---

[2]A system or process may have one or more operational modes which produce its output. In each operational mode, specific functions are performed that result in the output. Each function is directly related to one or more systems that perform the necessary functional actions. These systems, in turn, are composed of more basic units (e.g., components) that accomplish the objective of the system. As long as a system is operating within its design parameter tolerances, there is little chance of challenging the system boundaries in such a way that hazards will escape those boundaries. These operational modes are called normal operation modes.

During normal operation, loss of certain functions (e.g., protective functions) will cause the process to enter an off-normal condition. Once in this condition, there are two possibilities. First, the state of the process could be such that no other function is required to maintain the process in a safe condition. (*Safe* refers to a mode where the chance of uncontrolled exposure of hazards is negligible.) The second possibility is a state wherein other functions or systems are required to prevent exposure to hazards. For this second possibility, the loss of a function or loss of a system is an *initiating event.*

One method for determining the initiating events begins with drawing a functional diagram of the facility or process. From the functional diagram, a hierarchical relationship is produced, with the process objective successful completion of the desired process. Each function can then be decomposed into its subfunctions, systems, and human actions. Potential initiating events are the failures of particular functions, systems, or human actions. These potential initiating events are grouped such that members of a group have the same scenario of events for exposing the hazard.

required to respond following an initiating event.[3] The event tree fully delineates the process or system response to an initiating event and serves as the main tool for further analysis in the PRA.

The following procedures should be followed in this step of the PRA:

1. Identify the mitigating functions for each initiating event (or group of events).
2. Identify the corresponding human actions, systems or hardware operations associated with each function, along with their necessary conditions for success.
3. Develop a functional event tree for each initiating event (or group of events).
4. Develop a systemic event tree for each initiating event, delineating the success conditions, initiating event progression phenomena, and end effect of each scenario.

A more detailed discussion on how to develop event trees is presented in Modarres et al. (1999).

Build the Probabilistic Model of the System

The most common method used in PRA to calculate the probability of system failure is a fault tree analysis, a subset of event trees. A Boolean model is developed to quantify the contributions to risk by the process, equipment, human action, and inputs. For more reading on how to develop fault trees see Vesely et al. (1981) and Modarres et al. (1999).

The following procedures should be followed as a part of developing the fault tree:

1. Develop a fault tree for each event in the event tree heading.
2. Explicitly model dependencies between equipment, human operators, software, etc.

---

[3] In PRA, two types of event trees can be developed: functional and systemic. The functional event tree uses mitigating functions as its heading events. The main purpose of the functional tree is to better understand the scenario of events at a high level following the occurrence of an initiating event. The functional tree also guides the PRA analyst in the development of a more detailed systemic event tree. The systemic event tree reflects specific events (specific human actions or mitigative system operations or failures) that lead to a hazardous outcome. The functional event tree can be further decomposed to show specific hardware or human actions that perform the functions described in the functional event tree. Therefore, a systemic event tree fully delineates the process or system response to an initiating event and serves as the main tool for further analysis in the PRA.

3. Include all potential causes of failure, such as hardware, software, testing and maintenance, and human error, in the fault tree.

Events that originate within a facility or process are called internal events. In the context of the slaughterhouse, an example of an internal event might be failure of equipment to control microbial growth (i.e., a compressor failing in the chiller).

The clear counterpoint to an internal initiating event is an initiating event that originates outside of the facility, called an external event. Examples of external events are fires or floods that originate outside facility boundaries, seismic events, transportation events, volcanic events, high-wind events, and external epidemic conditions. Pathogens are often introduced to plants from the outside via contaminated animals. Thus, these events can be viewed as external events. Again, this classification can be used in grouping the event tree scenarios.[4]

To attain very low levels of risk, the systems, equipment, and human actions that comprise the barriers to hazard exposure must have very high levels of reliability. This high level of reliability is typically achieved through the use of redundant and/or diverse equipment and people, which provides multiple opportunities for successful hazard control. The problem then becomes one of ensuring the independence of the paths. Treatment of dependencies should be carefully included in both event tree and fault tree development and analysis in PRA. As the reliability of individual equipment or human action increases due to redundancy, the contribution from dependent failures becomes more important.[5]

---

[4]There is also a third category of events. Events that adversely affect the process and occur outside of the process boundaries are defined as internal events *external* to the facility. Typical internal events external to the process are internal fires and internal floods. The effects of these events should be modeled with new event trees to show all possible scenarios.

[5]In the nuclear industry, dependent failures can dominate the overall risk of exposure. Including the effects of dependent failures in the risk models (fault tree and event trees) is difficult and requires some sophisticated, fully integrated models. The treatment of dependent failures is not a single step performed during the PRA; it must be considered throughout the analysis (e.g., in event trees, fault trees, and human actions).

The following procedures should be followed in the dependent failure analysis:

1. Identify the items that are similar and could cause dependent or common cause failures. For example, similar equipment doing the same thing.
2. Items that are potentially susceptible to common cause failure should be explicitly incorporated into the fault trees and event trees where applicable.
3. Functional dependencies should be identified and explicitly modeled in the fault trees and event trees.

Data Analysis and Quantification

A critical building block in risk assessment is data on the performance of the equipment used to prevent and reduce hazards and the degree to which human actions influence hazards. In the slaughter plant, this includes knowledge of the operation and effectiveness of pathogen reduction technologies. In modeling pathogen propagation, we concentrate on data needed to establish the likelihood of mitigating and preventing hazards, as well as data on how frequently initiating events can occur. In particular, the best resources for predicting future effectiveness of equipment are past experiences and laboratory and *in situ* experiments. It must be recognized, however, that historical data have predictive value only to the extent that the conditions under which the data were generated remain applicable.[6]

The following procedures should be followed as part of the data analysis task:

1. Collect and statistically evaluate facility-specific data. Customize the failure probability distributions for control equipment and human actions in the fault-tree, event tree models.
2. If facility-specific data are not available, rely on generic sources such as industry-wide failure rates or data from another industry.[7] Assess and adjust the generic data to apply to the situation modeled.
3. Determine the frequency of initiating events and other failure events from experience, expert judgment, or generic sources.
4. Determine the common cause failure probability.

Application of Probabilistic Risk Assessment Methodology to a Beef Slaughterhouse

A PRA model explicitly details the scenario of events that must occur for hazards to transpire. Building a PRA model is a task that involves input

---

[6] In PRA for the nuclear industry, three types of events identified during the scenario definition and modeling must be quantified for the event trees and fault trees to estimate the frequency of occurrence of sequences: initiating events, failures of physical barriers, and human errors.

[7] Event trees commonly involve branch points at which a given piece of equipment or human operator either properly acts or fails to act. Sometimes, failure of the equipment or operator is rare or small and there may not be adequate historical records of such events to provide a dependable database. In this case, data from another industry or expert opinion can be used.

from many disciplines in a structured environment. The process used in developing the beef slaughterhouse model described in this chapter includes the steps outlined below:

- *Define scope of the problem* and identify the resources needed to accomplish the task.
- *Develop scenarios for an event tree* to describe the multiple sequence of events that can lead to contaminated product within the slaughter plant.
- *Build the probabilistic model* to reflect the plant-to-plant variability and uncertainty of plant practices in the beef slaughter industry.
- *Collect data,* including the issues surrounding the use of actual observed plant data vs. laboratory data vs. expert opinion, international vs. U.S. data and generic *E. coli* vs. *E. coli* O157:H7 data; evaluate the ramifications of using heterogeneous data sets.
- *Run Monte Carlo simulations* to obtain the distribution of contamination frequencies and levels for the target pathogen.
- *Validate and verify the model* to ensure that it accurately characterizes the behavior of the slaughter plant.
- *Perform sensitivity analysis* to rank risk-significant events to identify elements (human actions, events, and processes) that are important contributors to risk.
- *Combine the risk assessment model with economic information* to enable the evaluation of the cost effectiveness of alternative risk reduction options.

Description of the Process and Definition of the Scope

The slaughterhouse model is part of a larger, farm-to-table model of ground beef processing and consumption in the United States.[8] The slaughterhouse model begins with arrival of fed and cull cattle and ends with the production of combo bins (one-ton containers of the part of the beef carcass destined for ground beef production). The various major steps in the slaughterhouse are illustrated in Figure 1.1.

In the first step, cattle entering the slaughterhouse are stunned, hung from an overhead rail, bled, and their hides are removed. Carcass contamination can occur via contact with a pathogenic *E. coli*-contaminated hide/tail, gloves/clothing/hands, knives, or aerosols. Cross-contamination can also occur between animals or between carcasses.

[8] For background, see the USDA/FSIS *Preliminary Pathways and Data for a Risk Assessment of E. coli O157:H7 in Beef* on the web at *http://www.fsis.usda.gov/OPHS/ecolrisk/prelim.htm*.

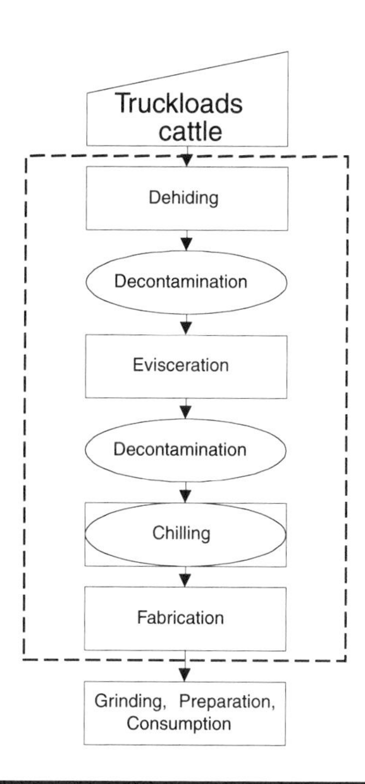

Figure 1.1 Major steps in the ground beef production process (boxes represent steps where contamination can occur, ovals represent steps where decontamination can occur)

Once the hide is removed, carcasses are trimmed and spot-steam vacuumed to remove visible contamination. During this process, pathogenic *E. coli* levels can be reduced or redistributed over the carcass. Next, carcasses are split open and eviscerated to remove the gastrointestinal (GI) tract, and sawed in half. During this operation, there is a possibility for the GI tract to rupture. If the animal producing this carcass was positive for *E. coli,* the carcass surface may become self-contaminated.

The carcasses proceed through a second decontamination process that can include one or more of the following operations: further steam vacuuming, hot water rinses of the carcass (perhaps in combination with organic acids or trisodium phosphate), or steam pasteurization of the carcass. During the carcass washes, *E. coli* can be killed, washed off, or redistributed over the carcass.

Carcasses are then hung in a refrigerated chamber and chilled for 18 to 48 hours. The amount of growth or decline of *E. coli* on the carcass surface is a function of time and temperature. Cross-contamination of a carcass by neighboring carcasses can occur in overcrowded chiller conditions. In the chiller, a carcass can experience either an increase or a decrease in the level of contamination.

The carcasses subsequently move to the cutting room, where they are fabricated to remove meat from the bones. Typically, the cut meat is packaged either in 60-pound boxes or 2000-pound combo bins. At this point there is potential for contamination from detritus on the equipment deposited by contaminated meat that arrived earlier on the fabrication line.

The cut meat is then transported, if necessary, for grinding into hamburger. Grinding may occur in the slaughter plant or at another facility. From the time the meat is packaged to the time it arrives at the grinder, there is a potential for growth of *E. coli* if the temperature within the truck and other storage facilities is above 45°F.

Understanding this process in sufficient detail to accomplish a PRA model is, for the most part, not common knowledge to the average modeler. Knowledge of practices inside the slaughterhouse is the domain of veterinarians, meat inspectors, and plant operators. PRA teams that do not have access to such information need to consult with individuals expert in the many facets of plant operations in the beef slaughter industry to make credible and reliable models.

Develop Scenarios for the Event Tree

The industry-wide generic event tree model is constructed for events that play a significant role in contamination or decontamination of beef carcasses in the slaughterhouse. To construct the scenarios, decision scientists and modelers are needed, working alongside subject matter experts to understand and capture the sequence of events in the plant that leads to measurable levels of contamination of beef.

Models can either be tailored to a specific plant or process or can be representative of a whole industry (generic event tree models). There are pros and cons to each approach. Plant-specific models are often easier to construct, and reliable data may be more readily available since a small set of personnel, management, equipment, and input variables are included. To compute a nationwide estimate of risk representing all beef slaughterhouses, variations among U.S. beef plants need to be considered. This could be done in several ways. Each plant in the U.S. (or region of interest) could be modeled separately and the results aggregated. This is a large-scale undertaking requiring detailed knowledge of the distribution of plant characteristics. Additionally, different classes of beef plants with close risk characterization could be modeled, and their risks combined. Finally, one representative model could be constructed that takes into account the variability in slaughterhouse practices throughout the U.S. Our model uses the final approach. An event tree corresponding to this representative model is illustrated in Figure 1.2.

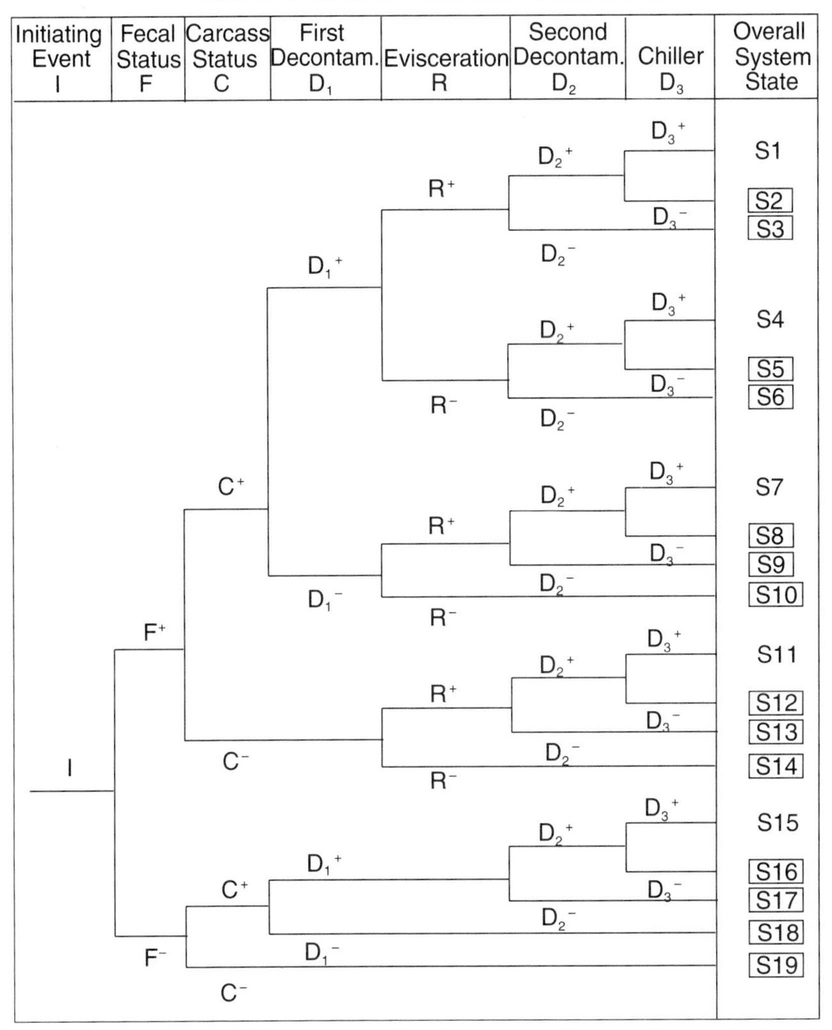

Figure 1.2 Event tree and possible outcomes in a slaughterhouse. □ = clean carcass status.

Building the Probabilistic Model

In this model we are concerned about the probability of a specific event occurring at each of a number of important steps. Table 1.1 describes some of the things that can go wrong in operating a slaughter plant, permitting contamination of the meat with a pathogen. The event tree shown in Figure 1.2 describes the steps in the slaughter process where these wrong events can happen. The input to the plant is live animals. An

Table 1.1 Things That Can Go Wrong in Operating a Slaughter Plant

Procedural factors

- Procedural failure because operating plan does not take target pathogen into account
- Outmoded procedures may be used that do not take advantage of latest scientific information on pathogen control
- Inadequate training plan for personnel in plant on pathogen control
- Management oversight plan is inadequate to control target pathogen
- Line speed and how it can alter processes and the probability of contamination or decontamination
- Recordkeeping or monitoring is inadequate to verify pathogen control in plant
- Lack of or inadequate performance bonus for personnel with good performance in pathogen control
- Microbial testing too limited to assure that plant sanitation is controlling pathogens

Equipment factors

- Performance problems with equipment from a particular supplier
- Performance problems with a particular piece of defective equipment
- Problems in maintaining equipment and replacing in a timely fashion at end of its life cycle
- Problems in how various types of equipment link together
- Extremely hot weather can strain chilling equipment inside plant

Input factors

- No testing of incoming cattle or too little testing to assure that incoming cattle are not contaminated
- Seasonal impact on contamination status of incoming cattle
- Possibility of other contaminated inputs

Personnel factors

- How work orders are transmitted or changed for each individual can lead to errors of commission or omission
- Examples of errors of commission
- Time for a worker to learn system and how to operate a particular piece of equipment
- Workers vary in how they operate a particular piece of equipment and some have higher probability of permitting contamination to occur
- All workers have good days and bad days when mistakes are made and things are done wrong
- Worker fails to do an action on a particular carcass and consequently there is a higher probability of pathogen contamination
- Worker fails to clean a particular piece of equipment at end of shift

Table 1.1 ***(continued)***

Miscellaneous factors
• Requirements by customers or state/local/federal regulatory bodies may change plant protocol and reduce pathogen control • Legal liability suits can have impact on plant operations and pathogen control (e.g., lawyers say don't test for pathogens) • Seasonal variability in product sales may strain pathogen control operations • Weekly variability in supply and demand (e.g., possible overflow of chiller on weekend) • Aerosol control within various parts of slaughter plant • Strain variability for *E. coli* O157:H7 and impact on decline, survival, or growth

incoming animal may be either fecal positive (F+) or fecal negative (F−).[9] When the hide is removed, the carcass can remain clean (C−) or the carcass can become contaminated with fecal matter (C+). At this point, the first decontamination procedure is applied. If the carcass is not contaminated, the procedure will not change the state of the carcass. Otherwise, the level of contamination will be reduced by some level, in some instances to zero (D_1). Upon evisceration, the GI tract of the carcass may rupture. If the animal is fecal positive, this may further contaminate the carcass (R). A second decontamination (D_2) is applied with consequences similar to the first. The carcasses are stored in a refrigerated chamber before being cut for further processing. In the chiller there can either be growth or decline of pathogens (D_3). In some rare instances, the level of contamination will be reduced to zero. The final outcome is either a clean or contaminated carcass. The states corresponding to a clean carcass are denoted by squares in Figure 1.2.

Data Collection

It is important that the model includes all of the relevant events so that it reflects the full range of possible outcomes. The specific data used should not affect the structural aspects of the model. Once a robust model has been built, data needed by the model is assembled. Once the data is in place, preliminary model runs can be made. We expect that the data collection will not be static, as the probabilities of contamination will change over time as processes and technologies improve.

[9] The + sign indicates that *E. coli* is found and the - sign indicates no *E. coli* found.

Many types of data are needed to model the beef slaughterhouse. Data include the probability that an event will cause meat to be contaminated with a pathogen (*E. coli* in this model), the surface area of meat contaminated, and the level of contamination over that area. It is in the realm of data collection that the specialized knowledge of domain experts is most vital. Animal microbiologists, industry experts, and reliability engineers each play a role in identifying sources of data and assessing their credibility and utility.

Data on actual plant operations is often difficult to obtain and its accuracy is sometimes questionable. The ability to use collected data to identify the probability and level of pathogen contamination is also problematic. Nevertheless, quality data plays a vital role in any modeling effort. When good data is available, it can be readily incorporated into models. As is often the case, however, the data availability and the modeling requirements often do not match. Using data that is less than ideal may be the only option. Indeed, data availability may drive some modeling decisions.

For our model, a hierarchy of data sources was established. Wherever possible, we used data obtained from plant-level studies carried out in the U.S. It was felt that such data would more accurately reflect the conditions we were modeling. If necessary, we used other sources such as non-U.S. studies, laboratory studies, and expert opinion. In all cases, we relied on experts to assess the value of the data.

Note that the choice of model type (specific plant versus representative plant) affects the data requirements and model variability and uncertainty. For our representative model, data is needed on practices in plants throughout the country as well as data specific to microbial behavior on the carcass surface, such as probability of contamination, area of the carcass affected, and level of contamination. An industry-wide representative model will include more variability and uncertainty because of the diverse practices among plants and our uncertainty about actual plant practices.[10]

Run the Monte Carlo Simulations

Risk assessment models in food safety have historically relied on mean values, (what happens in the average case), however, the use of mean values of risk does not take into account the range of risky outcomes. If beef cattle carcasses are contaminated at an average level of 2 log CFU[11]

---

[10] The uncertainty in an industry model can be converted to model variability by collecting more extensive data on actual plant practices and their impact on the probability of contamination and likely levels of contamination.

[11] CFU = colony-forming units and is a laboratory based method of counting the number of clusters of an organism that appear in a petri dish that has been inoculated with a sample.

per square centimeter (or 100 organisms/cm^2), and a decontamination process can remove an average of 90% of pathogens (1 log reduction), then the average contamination level of carcasses after decontamination is 1 log CFU (or 10 organisms/cm^2). In reality, the final contamination levels exemplify a range of values due to variability or uncertainty in plant practices. For example, a carcass with an initial 3 log CFU of contamination (above the average level) may only have 50% of pathogens removed (below the average level). Monte Carlo simulations provide a means to reflect the uncertainty and variability associated with estimates in a quantitative manner.

Monte Carlo simulations use probability distributions to describe the range of possibilities at each risk-increasing (contamination) or risk-decreasing (decontamination) step considering only the average or most likely values. The use of probability distributions is important, as it enables one to model both uncertainty in observed data and variability due to diversity of plant practices in the beef slaughter industry. In our beef slaughterhouse model (Figure 1.2), we used Monte Carlo simulations to capture the wide range of possible outcomes at each element in the event tree model.

Our modeling effort required the use of available data. We had neither time nor resources available to design experiments that would provide data specific to our needs. Where possible, actual data fitted to distributions. In other cases, probability distributions for the events were estimated from expert knowledge of industry practices.

Each uncertain and variable event in the model is assigned a probability distribution. Monte Carlo simulations are carried out by sampling from each event's distribution of data values. The model performs the calculations to determine the output for that sample. This process is repeated, usually hundreds to thousands of times, to create an output distribution of the probability and level of contamination associated with each scenario. In our slaughter model, the Monte Carlo simulations generated a distribution of probability of contamination in meat destined for hamburger. This distribution of contaminated meat is the principal output of our model. In the context of the overall farm-to-table model, it is also the input to the next module in the chain—the processing segment.

Validate and Verify the Model

Two questions naturally arise out of a model-building exercise: "Is the right model being built?" and "Is the model being built right?" These two questions may appear almost identical, but they are fundamentally different. The first question addresses the issue of validation of a model. Validation is achieved by checking the correspondence between reality and the output (including intermediate values) of the simulation. A model can only be said to be valid if it can reasonably predict the aggregate behavior of the system being simulated. If an adequate amount of data is not available for evaluation, expert opinion may be used. Validation is facilitated by periodic assessments by outside reviewers. Such reviewers might include industry and risk assessment experts who understand the PRA process.

The second question addresses verification. Verification concerns how the components are built and linked together. Each part of the model should be audited for accuracy and examined to insure conformity with theoretically sound procedures. Validation and verification are not just one-time steps. They are processes that must be carried out during the entire life cycle of the project.

Perform Sensitivity Analysis

Sensitivity analysis addresses how model output varies as conditions drive the output change. Such variation may reflect changes in data or in modeling assumptions. Sensitivity analysis thus provides a means to check the robustness of the output with respect to initial assessments and assumptions. By examining the degree to which important inputs affect the model output, we gain more insight and a better understanding of the simulation model.

Typically, a sensitivity analysis is performed by making changes in the input data, such as reducing or increasing the mean value of a distribution, and observing the resulting change in the output. In a large, realistic simulation model there may be dozens, if not hundreds, of parameters to consider. It is impractical to examine the sensitivity of all variables. One class of parameters that are good candidates for sensitivity analysis is those with the greatest level of uncertainty. These might include parameters based on small data samples. It is also useful to test the sensitivity of output to changes in model assumptions. For example, assumptions regarding worker behavior or equipment performance can be adjusted.

The results of the sensitivity analysis can point to areas where it is important to reduce uncertainty. If a small change in an uncertain parameter causes a disproportionately different model output, obtaining a more precise estimate of that parameter may be warranted. Of course, if uncertainty is an inherent feature of the parameter, then the model's users must accept or recognize that a high degree of uncertainty is a feature of the model output as well. If the output changes under two different assumptions surrounding a particular event, more thought is needed on which is the more appropriate.

In our beef slaughter model, the output of interest is the frequency of contaminated combo bins. We varied several important parameters by adjusting their means and variances both upwards and downwards, and observing the change in contamination level. We found that the range of contamination levels fell within our expectations and no single variable had a disproportionate impact on the output.

Integrate the Risk Model with Economic Information

The PRA model described above addresses the uncertainty and variability surrounding risk increasing and decreasing events in a quantitative manner. A useful next step of the analysis is to take the output of the PRA model and combine it with an economic analysis. The combination of these techniques enables us to evaluate the effectiveness of various technologies and their combinations. The method is straightforward. The PRA model is first run for the baseline case (i.e., no improved technologies are present), producing the cumulative distribution function (CDF), F_0. This CDF is shown in Figure 1.3. Further model runs are performed including one or more of the technologies to reduce contamination. For example, running the modified PRA model may produce the CDF described by F_1. This second distribution typically is shifted to the left of the first, reflecting the reduction in either pathogens or indicator organisms (see footnote 1). This shift reflects the degree to which contamination is reduced by including the technology.

From a risk assessment standpoint, we are interested in not simply the expected value of hamburger contamination, but rather the frequency with which some level of risk occurs in hamburgers. Focus for this risk assessment would be on the right of the underlying probability distribution rather than the mean value. To evaluate the effectiveness of technology

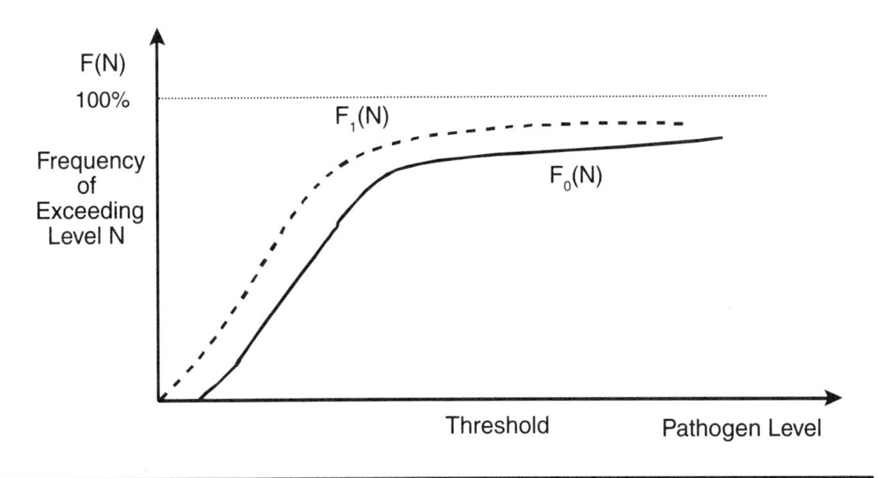

Figure 1.3 Cumulative distribution of *E. coli* levels in raw hamburgers

adoption strategies, we selected a risk tolerance threshold. By doing so, we expect that the change of expected pathogen frequency above the threshold compared to the baseline level represents effectiveness of the adoption strategy. This is expressed as:

$$\Delta P(\textit{hamburger contamination above threshold}) = (F_1(\textit{Threshold}) - F_0(\textit{Threshold}))$$

The difference $F_1 - F_0$ represents the change in the probability that a combo bin is above the risk threshold.

Next, the economic costs of the control strategies are combined with the PRA model results. Figure 1.4 shows how four hypothetical strategies (A, B, C, and D) might be compared. The x-axis represents the marginal cost of adding one of the new strategies compared to the baseline. The y-axis is the percentage reduction in contamination over the baseline. Strategy D can be excluded from any decision set, since strategy B dominates D in the sense that B is more effective and less costly. Choices of adoption strategies can be limited to non-dominated strategies A, B, and C. Each firm will have a unique optimal strategy curve at a particular point in time. The strategy a firm will choose depends on the risk preferences of their customers (Does the firm get a premium price for a safer product?) and on their comparative advantage in particular risk reduction strategies.

A plant's capability to utilize various technologies results in different adoption costs for similar technologies. A plant with a stable workforce may realize greater benefits from worker training, for example, in sanitary hide removal, because it has a lower likelihood of losing training value due to worker attrition than a plant with high worker turnover. Larger plants are more likely to profit from purchasing expensive new technologies that have significant economies of scale.

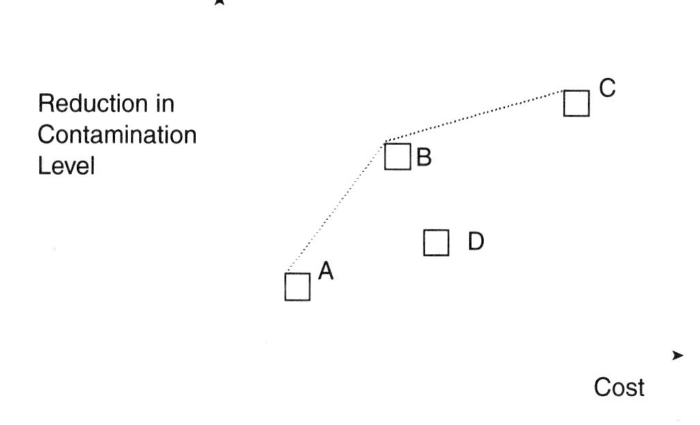

Figure 1.4 Comparison of adoption strategies

Both small and large plants may choose not to adopt some technologies, despite their effectiveness in pathogen reduction, because there are not the proper market incentives for adoption or for sufficient investment in research for development of new technologies.

Discussion

Risk assessment has evolved from straightforward identification of hazards to complex models incorporating probability distributions and uncertainty of knowledge (Rechard, 1999). By examining the models used in nuclear engineering and other disciplines, food safety risk assessors can learn about model structure and variables affecting reliability (Modarres, Krivtsor, and Kaminskiy, 1999). While modeling knowledge is necessary, it is not sufficient for building a food safety PRA model. Knowledge about the food industry and plant practices is also needed, so a food safety PRA team is wise to include food scientists, veterinarians, and economists as well as modelers.

To underscore the value of this aim, a step-by-step description of a process to build a PRA model for pathogen control was described. An example of an event tree was developed for a beef slaughterhouse. Each node in the tree indicates the possibility of contamination or decontamination that takes place as the animal/carcass moves through dehiding, decontamination, evisceration, a second decontamination, and chilling. Five of the nineteen paths result in contamination of the carcass. The level of risk presented by these five scenarios depends on the probabilities of the underlying events. Several strategies to minimize carcass contamination are possible: firms may decide to exert control evenly at all steps in the process, some may focus on preventing carcass contamination when the hide is removed, and others may focus on a comprehensive second decontamination to clean up any contamination that has occurred.

A previous paper (Roberts, Malcolm, and Narrod, 1999) used a simple model to evaluate four beef slaughterhouse interventions. In that model, careful hide removal was the most effective single strategy for reducing generic *E. coli*. However, as McDowell et al. state (1995: 120), "Food safety managers are faced with the problem of assembling a 'portfolio' of mitigation techniques to obtain some desired level of safety (or maximizing safety for a given cost)." We have addressed the portfolio issue by examining the risk–cost tradeoffs and by constructing an efficient frontier of risk-reducing strategies. Each firm has its own optimal decision, depending on its characteristics. Narrod et al. (1999) found that using combinations of technologies revealed synergies between different technologies.

Two issues impinge on the usefulness of PRA models to public and private policymakers: transparency and uncertainty. If the public, policymakers, and industry cannot understand the inputs and general structure of the model, they will have a hard time believing in the results. If the data in the model, or the models themselves, do not accurately represent real world slaughterhouse operations, the uncertainty inherent in the output will have reduced the model's relevance for public policymakers and private decision makers. Both concerns can be solved: the first by carefully explaining modeling assumptions and structure, the second by building a new collaborative relationship with industry in providing food safety databases and relying on advanced data analysis techniques such as the Bayesian approach (Modarres et al., 1999).

In conclusion, PRA models can make important contributions to food safety improvement programs, such as HACCP systems. HACCP is required of the meat and poultry industry by the USDA's Food Safety and Inspection Service, and the Food and Drug Administration (FDA) requires seafood suppliers to use HACCP. The FDA requires that juice producers either pasteurize their juice or use another process to reduce pathogens by 5 logs (for each 100,000 initial organisms, only 1 remains after the process). If neither treatment is used, the juice must contain a warning label: "WARNING: This product has not been pasteurized and, therefore, may contain harmful bacteria which can cause serious illness in children, the elderly, and persons with weakened immune systems" (p. 37043, *Federal Register*, July 8, 1998).

Other food companies have voluntarily implemented HACCP. HACCP suggests that control for complex systems should be concentrated in one or several places critical to the ultimate occurrence of hazard. By monitoring the control points carefully, the hazard's occurrence can be stopped or at least diminished. However, HACCP does not suggest a way of *identifying* the critical control points in a complex system. A well-developed PRA model can aid in this and other crucial decisions in managing the food chain for safety. By adding economic factors to the PRA model, risk-cost tradeoffs can be identified by permitting a benefit/cost analysis of alternative pathogen reduction strategies. Public and private decision makers can both benefit from using these PRA models combined with economic data.

References

Bisaillon, J.-R., Charlebois, R., Feltmate, T., and Labbe, Y., HACCP, Statistical Process Control applied to postmortem inspection and risk analysis in Canadian abbatoirs, *Dairy, Food, Environ. Sanit.*, 17, 150, 1997.

Cassin, M., Lammerding, A., Todd, E., Ross, W., and McColl, S., Quantitative risk assessment of *E. coli* O157:H7 in ground beef hamburgers, *Int. J. Food Microbiol.*, 41, 21, 1998.

Jensen, H., Unnevehr, L., and Gomez, M., Costs of improving food safety in the meat sector, *J. Agric. Appl. Econ.,* 30, 83, 1998.

McDowell, R., Kaplan, S., Ahl, A., and Roberts, T., Managing risks from foodborne microbial hazards, *Tracking Foodborne Pathogens From Farm to Table: Data Needs to Evaluate Control Options,* Economic Research Service/USDA, MP-1532, 1995, 117.

Modarres, M., Krivtsor, V., and Kaminskiy, K., *Reliability Engineering and Risk Analysis: A Practical Guide,* Marcel Dekker, N.Y., 1999.

Narrod, C., Malcolm, S., Ollinger, M., Roberts, T., Pathogen reduction options in slaughterhouses and methods for evaluating their effectiveness, paper presented at the American Association of Agricultural Economics Meetings, Nashville, Tennessee, 1999. Available on the web at http://agecon.lib.umn.edu/aaea99/sp99na04.pdf.

National Research Council, *Risk Assessment in the Federal Government: Managing the Process,* National Academy Press, Washington, D.C., 1983.

Rechard, R., Historical relationship between performance assessment for radioactive waste disposal and other types of risk assessment, *Risk Anal.,* 19, 763, 1999.

Roberts, T., Malcolm, S., and Narrod, C., Probabilistic risk assessment and slaughterhouse practices: modeling contamination process control in beef destined for hamburger, in *PSA '99: International Topical Meeting on Probabilistic Safety Assessment,* Modarres. M., Ed., American Nuclear Society, La Grange Park, IL, 1999, 809.

United States Department of Agriculture, Food Safety and Inspection Service, Preliminary pathways and data for a risk assessment of *E. coli* O157:H7 in beef, October 28, 1998. Available on the web at *http://www.fsis.usda.gov/OPHS/ecolrisk/prelim.htm.*

United States Department of Agriculture, Food Safety and Inspection Service, Pathogen Reduction; Hazard Analysis and Critical Control Points (HACCP) systems; final rule, *Federal Register,* Part II, 61(144), 38805, July 25, 1996.

United States Department of Health and Human Services, Food and Drug Administration, Food labeling: warning and notice statement, *Federal Register,* 63(103), 37029, July 8, 1998.

United States Department of Health and Human Services, Food and Drug Administration, Procedures for the safe and sanitary processing and importing of fish and fishery products: final rule, *Federal Register,* 60, 65197, December 18, 1995.

Vesely, W.E., Goldberg, G., Roberts, N., and Haasl, D., *Fault Tree Handbook,* NUREG-0492, U.S. Nuclear Regulatory Commission, Washington, D.C., 1981.

Chapter 2

Comparative Costs of Pathogen Reduction Strategies for Australian Beef Slaughter Plants

Vartguess Markarian, Neal H. Hooker, Elsa A. Murano, Gary R. Acuff, and Stephen Carroll

An interdisciplinary team was formed to determine effective and efficient pathogen reduction strategies for Australian beef slaughter plants. This evaluation takes the form of a comparative cost analysis of interventions designed to decontaminate carcasses. Laboratory-level, pilot-plant, and commercial facility evaluations are combined to provide the efficacy data. The cost data is collected from equipment supply companies and validated against current practices determined by a mail survey administered to beef slaughter plants in Australia. Key considerations have been found to include the scale of plants, cost of resources and labor, and export focus. Suggestions for the scale of societal benefits given the adoption of such interventions are forwarded. The lessons of the team and suggested future research complete the chapter.

Meat and Livestock Australia provided funding for this research.

0-8493-2217-0/01/$0.00+$.50

Introduction

An interdisciplinary team was formed to determine effective and efficient pathogen reduction strategies for Australian beef slaughter plants. This team brought together meat scientists, food microbiologists, veterinarians, and agricultural economists from the United States and Australia. The research team also called on support from various industry experts, government agencies, and most importantly, stakeholders throughout the beef supply chain. This chapter presents part of the research output of the team in the form of a comparative cost analysis of individual strategies and various combinations of strategies—so-called single and multiple-hurdle techniques–designed to decontaminate beef carcasses. These interventions include various cleaning (traditional trimming of visible contaminants, hand-held steam vacuums, and water rinses) and sanitizing (steam and acid cabinets, and hot water rinses) regimes. Information from the meat science literature, laboratory-level evaluations of various interventions, pilot-plant tests, and commercial facility experiences are combined to provide the efficacy data. The cost data was constructed from a review of recent food safety economics research and information collected from key equipment supply companies. This information was validated against current practices derived from a combination of on-site interviews, questionnaires, and a mail survey instrument administered to beef slaughter plants in Australia.

Considerable progress in the economics of food safety has been made over the last decade, with many applications focusing on the development of the United States Department of Agriculture (USDA) *Pathogen Reduction; Hazard Analysis and Critical Control Point (HACCP) Systems; Final Rule* (USDA 1996). Possibly the largest single global regulatory reform in the meat and poultry sector, the rule required plants supplying the U.S. market to implement HACCP-based quality assurance (QA) systems. While not a requirement, many plants in the U.S. and elsewhere have taken this opportunity to revise their processes and implement additional pathogen reduction strategies. Together with a greater availability of novel interventions (e.g., steam pasteurization) and strong customer demands for such process modification, there is a need for plants to determine which investments are most appropriate given scale, resource costs, and market requirements.

This chapter focuses on the direct additional fixed and variable costs and microbiological benefits of each intervention. Key direct costs include the initial capital (non-recurring cost), maintenance, training, labor, and inputs (recurring costs) solely due to the strategy under consideration (e.g., steam vacuum equipment). Food safety benefits arise primarily from safer food and are evident at the societal (public) level. To indicate these, we report particular plant level improvements represented by microbiological $\log_{10}$ reductions. Following the adoption of the various strategies, we assume that these plant level improvements can be aggregated to the public level. Additional firm level (private) benefits are also briefly discussed.

Experimental Design

Comparative Cost Analysis

A comprehensive review of the available literature quickly indicated that no single source or combination of sources could provide the information necessary to conduct an up-to-date consistent and accurate comparative cost analysis of pathogen reduction strategies for beef carcasses. Indeed, the research team was not able to identify a strong enough description of current industry practices as they relate to microbiological food safety issues, or even of industry structure. Therefore, a comprehensive industry survey was undertaken to ensure that accurate production assumptions for the costs could be made. The methodology for the development, administration, and analysis of this survey is discussed below.

An exhaustive survey development process was adopted to ensure that all issues and concerns of stakeholders were incorporated. This process involved pre-testing and review of a draft instrument by the research team, plant managers, and industry experts. The final mail survey instrument included 54 questions divided into 4 sections. Respondents were asked to indicate current pathogen control strategies used, costs of implementation, and their impacts on the whole slaughter process. The second section focused on the QA systems in use, their implementation costs and in-plant effects, and impact on supplier/customer relations. The third section concerned identity preservation (the ability to "traceback" a product's production and processing history), while the final section collected production indicators such as throughput, sales, capacity, etc.

The frame from which the survey sample population was selected was based on various industry sources, making a special effort to include all major beef slaughter plants. A final sample of 98 plants was selected from the population of organizations comprising the Australian beef processing industry. Given the relatively small number of plants available, each processor was selected to maximize the likelihood of response while still retaining sample diversity in terms of plant location, throughput, and market orientation. Prior to mailing the survey, the management of each plant was contacted and informed of the background and objectives of the project, and asked whether they were willing to participate in the research. Following their approval, the survey was mailed accompanied by a comprehensive cover letter detailing the background of the project and researchers, project objectives, and security provisions ensuring anonymity of responses. The participating companies were given three weeks to respond, during which time two follow-up calls were made to motivate respondents, gauge their progress, and address any inquiries.

Given the complex nature of the livestock-beef supply chain and the variety of firms involved, the research team made a special effort to meet with individuals knowledgeable about as many types of operations as possible to augment the data collected by the survey. This included discussions with various industry associations, government agencies and research bodies, and supermarket buyers. Each of these discussions provided valuable information that has been incorporated into this chapter. Finally, and of great importance to the research, the team met with and/or telephoned each of the key equipment supply companies to ensure that accurate product information was included.

Findings: Results of the Survey Questionnaire

A considered and structured approach to the administration of the survey contributed to a relatively high response rate of 46% (45 plants). Of these, 20 were export registered and 25 were domestic plants.

Current Pathogen Control Strategies

Several important differences between domestic and export plants that can influence the selection or relevance of a particular pathogen reduction strategy were discovered. While the majority of plants in both categories use a particular hide removal and rail system, the popularity of each differs according to market orientation. Over 80% of export plants use a downward hide puller, as opposed to only 24% of domestic plants. This scenario is replicated in the case of rail systems. The majority (75%) of export plants indicated having a moving chain system, whereas nearly three-quarters of domestic plants use a gravity chain. While none of the interventions discussed in this report require an upward or downward hide puller, the rail system can be a factor. The predominance of gravity rails in domestic plants may suggest the relevance of certain interventions (e.g., steam vacuums or hand-held rinses) that would not require a substantial change to the rail system.

Steam is readily available in the majority of plants. However, more often than not, smaller domestic operations lack rendering plants and/or steam, again narrowing the range of options or requiring additional costs to introduce certain interventions (steam generation units cost between $4,000 and $14,000 depending upon size required)[1]. The cost of steam generation units has not been included explicitly; alternatively a per-ton cost of steam is assumed (see below).

Production Indicators

As expected, the average number of fed cattle slaughtered in export plants (349 grass fed, 168 grain fed) was far greater than in domestic plants (104

---

[1]All values are reported in Australian dollars. An approximate exchange of (A) $1.64 = (US) $1.00 can be used to convert.

grass, 84 grain). Similar differences were seen in the average weight of these fed cattle (over 300 kg live weight for export plants versus approximately 200 kg live weight for domestic plants). All but two domestic plants are multispecies, mostly slaughtering pigs and sheep, whereas this trend is reversed for export plants, with only two plants indicating that they slaughter other animals. Specialization in slaughter operations is one trend that must be considered during major process modification and capital investment. The applicability of these pathogen reduction strategies to other species has not been discussed in this chapter, but is clearly a concern of domestic plants.

The larger scale of export-registered plants can be reflected in the number of staff. The average number of fulltime slaughter, boning room, and rendering staff in export plants (76, 116, and 5, respectively) far exceed those in domestic (41, 13, and 2). The numbers of part time workers in each type of plant are similar. Whereas both types of plants averaged 5-day work weeks, export plants tended to run two 8-hour shifts, compared to domestic plants which run single 7-hour shifts.

Like employee numbers, a comparison of average annual sales reflects the generally larger throughput of export plants. In 1998, average annual sales by export plants fell between $100 to 199 million, compared with the range of $10 to 49 million suggested by the majority of domestic plants. Similarly, the value of plants and ability to invest significant capital obviously differs between plant types.

Products sourced from export plants are primarily in the form of quarters, boxed beef, primal cuts, and offal, whereas in domestic plants sides also feature prominently. Furthermore, 96% of domestic product is chilled, while this proportion declines to 50% in the case of export plants. If the trend of increasing value adding prior to export continues, the need to better understand the link between carcass microbiological levels following each intervention and levels on cuts should be further analyzed. There is not a large discrepancy between the average number of customers serviced. However, the figures suggest that export plants service a relatively small number of high-volume customers primarily on a fixed volume contract basis. With an average of 88 customers, domestic plants service about 20 more customers than export plants on a primarily service kill basis.

Results and Discussion

Findings: Constructing Cost Estimates

The cost of each pathogen reduction strategy is estimated based on a number of important assumptions relating to plant throughput, equipment specification, and resource prices. These assumptions are discussed below. Given the survey data complemented with additional validation exercises conducted during the research project, three size categories (small, medium, and large) were selected as presented in Table 2.1. Scale has been assigned based on the average hourly throughput (head of cattle per hour) for each plant, which provides a good reflection of equipment needs. For the purposes of cost calculations, a single point estimate (shown in parentheses) has been assumed for each category. It is assumed that large export plants operate two shifts, while the remaining export and domestic plants operate a single shift. It is further assumed that the average shift length of export plants in all size categories is eight hours, compared with seven hours in the case of domestic plants. All plants are assumed to operate five days per week for fifty weeks per year. Average live weight of cattle slaughtered in export and domestic plants is assumed to be 320 and 200 kg, respectively.

The operating cost of each pathogen reduction strategy is estimated in terms of its component parts, e.g., water, labor, etc. Optimal operation of each strategy will require different levels of resources and hence will yield a particular marginal cost of decontamination. The assumed resource requirements associated with each of the strategies are discussed below.

Trimming

The number of dedicated trimming stations in a plant is influenced by the management plan in place. A number of survey respondents indicated that most slaughter personnel are required to undertake trimming when/where required. However, the figures presented in Table 2.1 are assumed to reflect a broad range of observations of dedicated trimming stations across export and domestic plants of varying sizes. For the purposes of cost calculation, a single point estimate (shown in parentheses) has been assumed for each plant category.

Cold Water Wash

Automatic carcass wash systems can consume 50 to 120 kl/day. However, many plants have replaced such systems with a manual spray wash that is used when necessary. The manual spray is operated by a trimmer and hence adds a labor requirement to cold water washing activities. It is assumed that 25% of the trimmer's time will be used to perform washing tasks.

Hot Water Wash

The major costs associated with the hot water treatment of beef carcasses include water, electricity, steam, effluent disposal, and labor during the

Table 2.1 Scale Assumptions and Resource Costs

	Small	*Export Medium*	*Large*	*Small*	*Domestic Medium*	*Large*
Throughput (head/hour)[a]	<40 (20)	40 – 100 (70)	>100 (125)	<20 (10)	20 – 50 (35)	>50 (60)
Number of Shifts	1	1	2	1	1	1
Shift Length (hours)	8	8	8	7	7	7
Number of Trimming Stations[a]	<3 – 5 (3)	3 – 8 (5)	7 – >8 (8)	0 – 1 (1)	1 – 2 (1)	1 – >2 (2)
			Resource Costs			
	Labor	$16.50/hr		Electricity	$0.085/kWhr	
	Water	$0.50/kl		Steam	$15.00/ton	

a. Numbers in parentheses indicate point estimates used for cost estimates.

clean-down phase. Water, steam, and electricity requirements vary considerably, with throughput from 2500 l/hr, 375 Kg/hr, and 32.5 kW/hr for 50 head/hr to 7815 l/hr, 1875 kg/hr, and 50 kW/hr for 250 head/hr. Negligible labor time is required.

Steam Vacuum

The major resource requirements of the steam vacuum include labor: equivalent to one trimmer; steam: 12 kg/hr at 150°C/120°C × 50 p.s.i.; and water: 4–5 l/min at 82°C × 50 p.s.i. Based on manufacturers' literature describing the optimal throughput/unit ratio, it is assumed in this study that plants which have a daily throughput of 500 to 750 head will install three units, while those falling under that threshold will ideally install two units.

Steam Pasteurization

The most appropriate system for the Australian market has an optimal throughput of up to 100 bodies/hr, steam requirements of 10 kg/head, and water 60 l/hr.

Organic Acid Rinses

Labor costs are assumed to be approximately 1 hour/day, and water 5 l/head.

Input Costs

The variable cost components of trimming include labor, knives, and the opportunity cost of loss of weight. The cost of labor in the trimming function varies from $11.50 to $12.50/hr of actual work. Allowing for on-costs ranging from 38 to 50%, the labor cost of trimming increases to $15.90 to $18.75. This cost analysis will assume a loaded labor cost of $16.50/hr.

Water costs vary widely across Australia, from the low cost associated with simply recovering and chlorinating bore water to the high per kiloliter costs incurred in Western Australia. The observed range is from single cents per kiloliter to $1.60 per kiloliter. The modal figure of $0.50 is assumed in this analysis.

Although the cost of sawdust and coal is about one-third of gas costs, this low fuel cost is countered by the high capital, operating, and maintenance costs associated with these sources of steam generation, leading to a range of $8.00 to $15.00/ton of steam. The cost of gas-generated steam is likely to be in the order of $15.00 to $20.00/ton of steam. Steam is only generally available in plants that have associated rendering facilities, as meat-processing plants only require hot water. For this analysis, however, it is assumed that direct steam heating will be used at a cost of $15.00/ton of steam.

The total cost of electricity ranges from around $0.06 to $0.11/kWhr. In recent times, electricity costs have fallen due to the arrival of privatization and contestability. It is currently predicted that electricity prices are on the rise, based on evidence from recently completed contracts. For the purposes of this analysis, an electricity cost of $0.085/kWhr has been assumed. The cost of organic acid is assumed to be $0.30/carcass. Finally, it is assumed that all pieces of decontamination equipment depreciate at a rate of 10% per annum.

Given these resource costs and scale assumptions, it is possible to construct estimates of the fixed and variable costs of the individual pathogen-reduction strategies for each plant category. These estimates are presented in Tables 2.2 and 2.3 on a per hour and per kg live weight basis. While it is interesting to review these interventions in isolation, a significant contribution of the research conducted within this project was the comparison of various multiple-hurdle pathogen reduction strategies. The combination of an intervention designed to clean the carcass followed by a sanitizing intervention has most frequently been proven to be most efficacious. When considering the related costs of these multiple-hurdle interventions, costs may be additive (e.g., steam vacuum followed by lactic acid rinse). Alternatively, there may be cost savings in the multiple-hurdle strategy adopted. For example, a pre-wash may be followed by a hot water rinse that can be applied in a single cabinet. However, such returns to scope in pathogen reduction strategies are rare; therefore for ease, costs are simply aggregated.

Findings: Ranges of Pathogen Reductions

The microbiological efficacy figures for the various cleaning and sanitizing interventions both independently and in certain combinations were collected during an earlier phase of the project summarized in this chapter. Laboratory-level evaluations of the interventions, pilot-plant tests, and commercial facility experiences were combined to provide this data.[2] These efficacy figures can be presented in terms of the $\log_{10}$ reductions in pathogenic or indicator organisms. A subset of this information is

---

[2]Further details are available from the authors upon request.

Table 2.2 Fixed, Variable, and Total Costs of Individual Interventions—Export Plants (per hour)

	Intervention	*Trim*	*Cold Water Washes (small)*	*Cold Water Wash (large)*	*Steam Vacuum*	*Hot Water Flush*	*Lactic Acid Rinses*	*Steam Pasteurization*
Fixed Costs								
	S	(na)	15,000	15,000	52,350	330,000	150,000	550,000
	M	(na)	15,000	15,000	58,350	330,000	150,000	550,000
	L	(na)	15,000	15,000	58,350	338,000	150,000	550,000
Variable Costs								
	S	49.50	1.40	3.4	34.43	3.86	6.07	3.01
		(.31)	(.07)	(.17)	(1.72)	(.19)	(.30)	(.15)
	M	82.50	3.50	8.4	51.65	13.51	21.81	10.54
		(.15)	(.05)	(.12)	(0.74)	(0.19)	(.30)	(.15)
	L	132.00	3.75	7.5	51.65	20.13	37.75	18.81
		(.07)	(.03)	(.06)	(.41)	(.16)	(.30)	(.15)
Total Cost								
	S	49.50	2.15	4.15	37.05	20.36	13.57	30.51
		(.31)	(.11)	(.21)	(1.85)	(1.02)	(.68)	(1.53)
	M	82.50	4.25	9.15	54.56	30.01	28.68	38.04
		(.15)	(.06)	(.13)	(.78)	(0.43)	(.41)	(.54)
	L	132.00	4.13	7.88	53.10	28.38	41.50	32.56
		(.07)	(.03)	(.06)	(.41)	(.023)	(.33)	(.26)
Total Cost/kg (l.w.)								
	S	.010	>.001	.001	.006	.003	.002	.005
	M	.001	>.001	>.001	.002	.001	.001	.002
	L	>.001	>.001	>.001	.001	.001	.001	.001

Note: Figures in parentheses indicate costs on a per head basis live weight (l.w.).

Table 2.3 Fixed, Variable, and Total Costs of Individual Interventions—Domestic Plants (per hour)

	Intervention	*Trim*	*Cold Water Wash (small)*	*Cold Water Wash (large)*	*Steam Vacuum*	*Hot Water Flush*	*Lactic Acid Rinse*	*Steam Pasteurization*
Fixed Costs								
	S	(na)	15,000	15,000	52,350	330,000	150,000	550,000
	M	(na)	15,000	15,000	52,350	330,000	150,000	550,000
	L	(na)	15,000	15,000	52,350	338,000	150,000	550,000
Variable Costs								
	S	16.50	.70	1.7	34.43	1.93	3.13	1.51
		(1.65)	(.07)	(.17)	(3.44)	(.19)	(.31)	(.15)
	M	16.50	2.45	5.95	34.43	6.76	10.79	5.27
		(0.47)	(.07)	(.17)	(.98)	(.19)	(.31)	(.15)
	L	33.00	4.20	10.2	34.43	11.58	18.16	9.03
		(0.55)	(.07)	(.17)	(.57)	(.19)	(.30)	(.15)
Total Cost								
	S	16.50	1.56	2.56	37.42	29.79	11.70	32.93
		(1.65)	(.16)	(.26)	(3.74)	(2.08)	(1.17)	(3.29)
	M	16.50	3.31	6.81	37.42	25.61	19.36	36.70
		(0.47)	(.09)	(.19)	(1.07)	(.73)	(0.55)	(1.05)
	L	33.00	5.06	11.06	37.42	30.44	26.73	40.46
		(0.55)	(.08)	(.18)	(.62)	(.51)	(.45)	(.67)
Total Cost/kg (l.w.)								
	S	.008	.001	.001	.019	.010	.006	.016
	M	.002	.001	.001	.005	.004	.003	.005
	L	.003	.001	>.001	.003	.003	.002	.003

Note: Figures in parentheses indicate costs on a per head basis live weight (l.w.).

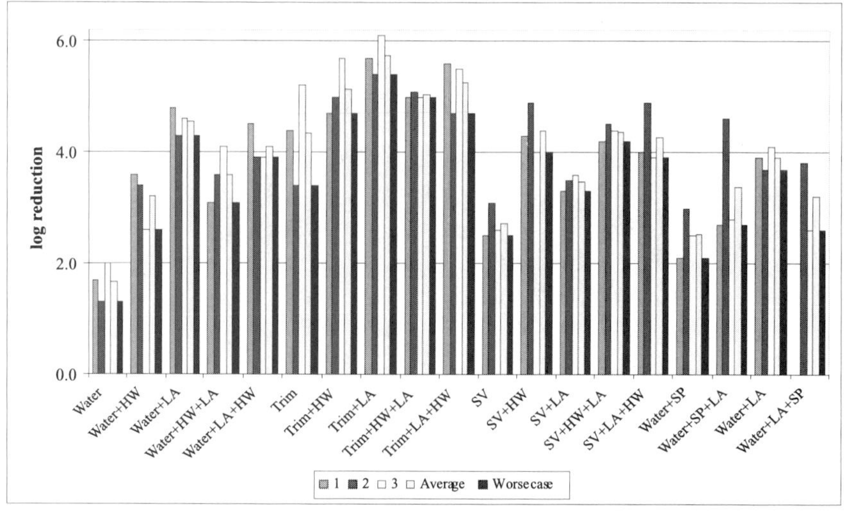

Figure 2.1 Total Active Plate Count Log_{10} Reductions for Single and Multiple Hurdle Interventions

presented in Figure 2.1, including individual $\log_{10}$ reductions in the total plate count numbers following each intervention.[3] Average and minimal values for each intervention are also plotted.

Combining the Cost and Microbiological Data

It is not possible to present detailed cost and pathogen reduction estimates for every particular plant. Alternatively, the methodology applied here uses the most likely costs and reductions based on the notion of average export and domestic plants constructed above and the previously discussed microbiological efficacy data. One way to present this information that captures the essence of the problem is presented in Figure 2.2 (based on the work of Jensen et al., 1998; Narrod et al., 1999; and Roberts et al., 1999; also discussed in the Roberts et al. chapter in this volume). The clear nature of the trade-off between additional costs and increased levels of pathogen reduction over four possible intervention strategies (A, B, C, and D) is highlighted. Strategies such as D are dominated by other strategies with either higher log reductions (C), lower costs (A), or some ratio of lower costs and higher log reductions (B). The outer envelope (line) of

[3]The total number of organisms (bacteria, yeasts, and molds) present in a food sample following a pathogen reduction strategy compared to a control samples.

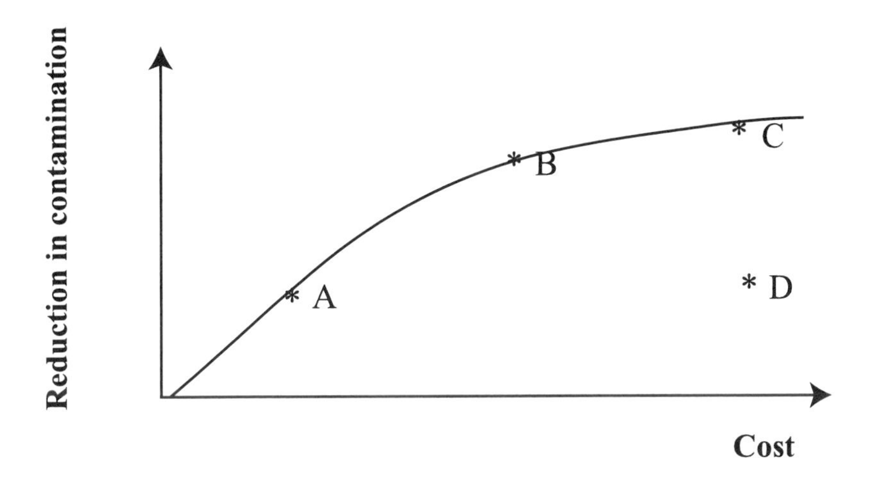

Figure 2.2 Trade-off of Pathogen Reductions and Costs

strategies marks the feasible and efficient mix of interventions. However, as information on only a limited set of pathogen reduction strategies exists, this is not to be read as an infinite set of alternatives, rather as a frontier that marks efficient combinations of cost and efficacy data.

The cost and efficacy trade-off curves for small, medium, and large domestic and export plants over the range of interventions evaluated here are presented in Figures 2.3 and 2.4. This presentation is the first assessment of the importance of scale and market focus in determining costs for

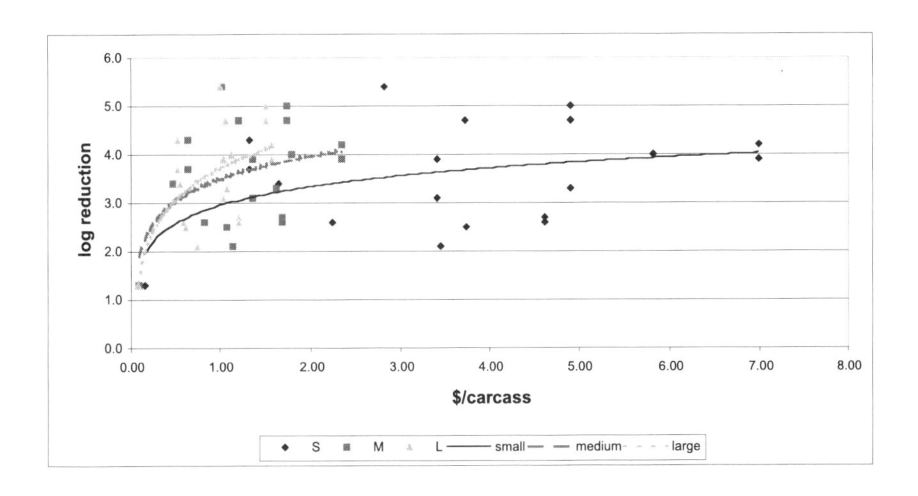

Figure 2.3 Total Cost and Total Plate Count Log_{10} Reductions Trade-off Curves – Domestic Plants

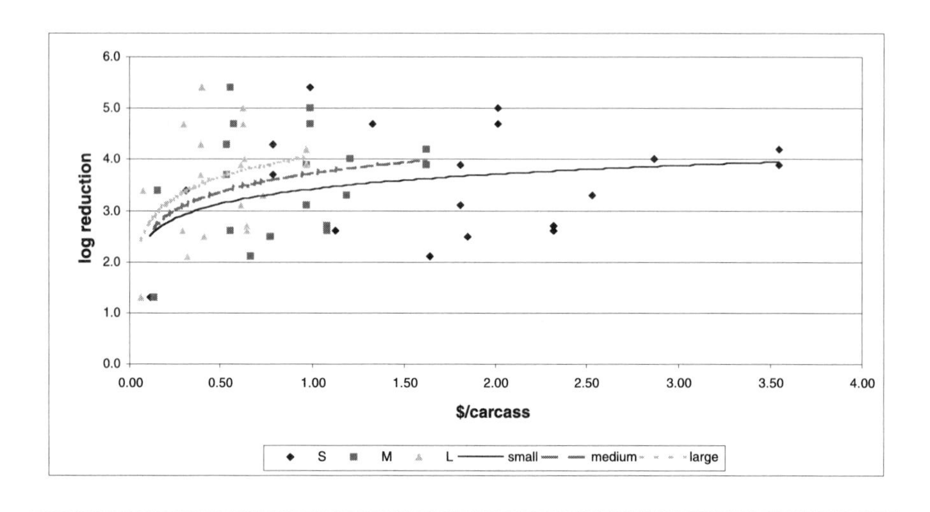

Figure 2.4 Total Cost and Total Plate Count Log_{10} Reductions Trade-off Curves – Export Plants

pathogen reduction strategies. No obvious difference was determined between the various indicators of microbiological efficacy employed (i.e., similar plots exist for total coliforms, generic *Escherichia coli,* etc.). Thus interventions may potentially be presumed to be equally effective regardless of the target organism. Certain basic relationships as demonstrated by the trade-off curves are evident. Economies of scale in pathogen reduction are obvious, with large plants frequently averaging per carcass costs one third of those for smaller operations. Further, comparing the two sets of plots demonstrates that export plants enjoy a cost advantage in implementing pathogen reduction strategies, simply through their ability to divide significant fixed costs over larger production volumes.

Clearly, any comparative cost analysis is influenced by the assumptions employed. A form of sensitivity analysis of these assumptions can be conducted using cost minimization. Given a target $\log_{10}$ reduction as the criteria for selecting an intervention, it is useful to know which intervention in isolation or in combination provides this efficacy at the lowest cost. Table 2.4 presents a sample of such information, indicating again the trade-off of increased cost and enhanced microbiological food safety control.

Expanding the Model

This simplified model fails to acknowledge the uncertainty and variability inherent in the adoption of particular strategies within and between plants. These dynamics can greatly influence both the cost of adoption of a particular intervention over a range of plants and its effectiveness in reducing

Table 2.4 Total Cost Matrix of Target Log Reductions ($/carcass)

Target Log Reduction —APC	Strategy	Domestic Plants			Export Plants		
		S	M	L	S	M	L
1.0 < x = <**2.0**	W	0.16	0.09	0.08	0.11	0.13	0.06
2.0 < x = <**3.0**	W + HW	2.24	0.82	0.59	1.13	0.56	0.29
	SV	3.74	1.07	0.62	1.85	0.78	0.41
	W + SP	3.45	1.14	0.75	1.64	0.67	0.32
	W + SP + LA	4.62	1.69	1.20	2.32	1.08	0.65
	W + LA + SP	4.62	1.69	1.20	2.32	1.08	0.65
3.0 < x = <**4.0**	W + HW + LA	3.41	1.37	1.04	1.81	0.97	0.62
	W + LA + HW	3.41	1.37	1.04	1.81	0.97	0.62
	T	1.65	1.37	1.04	0.31	0.15	0.07
	SV + LA	4.91	0.47	0.55	2.53	1.19	0.74
	SV + LA + HW	6.99	1.62	1.07	3.55	1.62	0.97
	W + LA	1.33	2.35	1.58	0.79	0.54	0.39
	SV + HW	1.13	0.64	0.53	2.87	1.21	0.64
4.0 < x = <**5.0**	W + LA	1.33	0.64	0.53	0.79	0.54	0.39
	T + HW	3.73	1.20	1.06	1.33	0.58	0.30
	T + LA + HW	4.90	1.75	1.51	2.01	0.99	0.63
	SV + HW + LA	6.99	2.35	1.58	3.55	1.62	0.97
	T + HW + LA	4.90	1.75	1.51	2.01	0.99	0.63
5.0 < x = <**6.0**	T + LA	2.82	1.02	1.00	0.99	0.56	0.40

the microbiological load on the product. A more advanced methodology for addressing this weakness has recently been proposed by Narrod et al., 1999, and an extended discussion is available in Roberts et al., 1999 and in Chapter 1 of this book. Probabilistic risk assessment tools are used to simulate potential distributions of effectiveness for a range of pathogen reduction strategies in cow and steer/heifer slaughter plants in the U.S.

In the construct of Figure 2.2, this can be considered as suggesting a range and distribution of $\log_{10}$ reductions likely to be observed for each strategy rather than the point estimates (worse case scenarios) used here. The efficacy of a particular intervention is then allowed to vary between plant types. The outer envelope of efficient strategies can then be thought of as marking the modal points of each distribution for all possible sets of interventions. One needs to carefully consider the nature of the underlying distributions in such assessments, for an incorrectly specified distribution (e.g., normal, when asymmetric may be more appropriate) can significantly bias the results in favor of one intervention or another. Essentially, the probabilistic risk assessment methodology assumes certain distributions for key variables (such as the $\log_{10}$ reduction observed by a particular type of plant) and then performs a high number of Monte Carlo simulations to construct a final distribution on the overall effectiveness of the system. The ability to then compare the broader structural impacts of adoption rates for each intervention and the potential reduction in the prevalence of foodborne illness motivate these recent theoretical advances.

A recent report released by Australia New Zealand Food Authority (ANZFA 1999) is useful in demonstrating the potential public health benefits of the adoption of pathogen reduction interventions in beef slaughter plants. It has been estimated that 11,500 people suffer some form of foodborne illness in Australia each day, for a total of 4.2 million cases per year. This conservative number translates to some $2.6 billion in costs per year borne by consumers, industry, and the public sector. Obviously, not all of these illnesses are due to beef, and the introduction of any intervention cannot eliminate all foodborne illnesses from any particular food source. Alternatively, an estimate of the potential societal impact can be made assuming certain criteria. In the U.S., for example, it has been suggested that approximately 6% of all foodborne illness outbreaks can be traced to the consumption of beef. If a similar percentage carries over to Australia, this would suggest some 252,000 cases and $156 million per year. This simple linear transformation fails to recognize that many meat-related foodborne illnesses result in far greater costs on average. These figures presume a standard cost per illness of some $315 to consumers, and $315 divided between the affected firm or industry and the public sector (ANZFA 1999, p.38).

Clearly there are other potential private benefits of pathogen reduction interventions. There may be "joint products" observable when targeting microbiological hazards, such as an enhanced ability to address physical and chemical hazards via the intervention. Further, shelf-life benefits have been observed following certain interventions, though this area of research remains ripe for additional focus. Other hypothesized, though difficult to quantify, benefits for enhanced pathogen reduction include tighter process control (e.g., reduced wastage), better management and worker responsibilities, reduced insurance premium costs, as well as stable and even expanded markets (domestic or export) following increased levels of trust by key customers.

Lessons Learned and Future Research

The developing field of food safety economics can still benefit from a wide discussion in Australia. It is hoped that this chapter will encourage further analysis of the costs and benefits of public and private strategies to enhance the safety of the Australian food supply. While this comparative cost analysis marks an early step in the selection of a particular intervention by a plant, it is hoped that the potential impacts on average domestic and export plants have been highlighted. A range of individual and multiple-hurdle pathogen reduction strategies is assessed for cost and ability to reduce the microbiological contamination level of beef carcasses. The methodology used in this chapter suggests how microbiological and cost data can be combined in both a visual and simple matrix form to demonstrate the trade-off of costs and benefits for various pathogen reduction strategies for beef slaughter plants in Australia.

This research provides a detailed extension of the material presented in Chapter 1. Information specific to the problem addressed (i.e., beef carcass pathogen reduction strategies that are relevant for Australian plants) was collected and then interpreted by an interdisciplinary, international team during the course of the research project. The team found that regular structured progress reports required by the funding body facilitated team discussions, forcing the various disciplines to keep up to date with the project's progress. Team members found the challenge of understanding each other's disciplines and contributions frustrating at times, but were open to new ideas throughout.

In the future it is hoped that as more detailed data become available, this research can be both updated and expanded. As with any risk assessment and cost effectiveness study, the data entering the model controls the quality of the results. The team feels strongly that the technique of combining microbiological and cost data adopted here can easily be adapted and expanded to assist plant management in selecting the most appropriate intervention, risk managers, and policy analysts in determining the public health benefit of food safety risk reductions.

References

Australia New Zealand Food Authority (ANZFA), *Food Safety Standards—Costs and Benefits: An Analysis of the Regulatory Impact of the Proposed National Food Safety Reforms,* May, 1999.

Jensen, H.H., Unnevehr, L.J., and Gomez, M.I., Costs of improving food safety in the meat sector, *J. Agric. Appl. Econ.,* 30(1), 83–94, 1998.

Narrod, C.A., Malcolm, S.A., Ollinger, M., and Roberts, T., Pathogen reduction options in slaughterhouses and methods for evaluating their economic effectiveness, paper presented in American Agricultural Economics Association meetings, Nashville, August, 1999.

Roberts, T., Malcolm, S., and Narrod, C.A., Probabilistic risk assessment and slaughterhouse practices: modeling contamination process control in beef destined for hamburger, paper presented at the International Topical Meeting on Probabilistic Safety Assessment Meetings, Washington, D.C., August, 1999.

U.S. Department of Agriculture, Pathogen Reduction; Hazard Analysis and Critical Control Point (HACCP) Systems; Final Rule, *Federal Register,* 61(144), 38805–38989, 1996.

Chapter 3

Institutional Arrangements and Incentive Structures for Food Safety and Quality Assurance in the Food Chain

Jill E. Hobbs, John Spriggs, and Andrew Fearne

Different public and private approaches to regulating and assuring food safety in three countries are compared: the U.K., Canada, and Australia. Although there are a number of similarities in the development of food safety initiatives in the three countries, there are also some important differences in key drivers for change and their impact. Key differences include the incidence of food safety scares, differences in incentive structures, and the proliferation of standards. Lessons from economic, management, and marketing literature shed light on the incentives for change from both a public policy and private industry strategy perspective. Vertical industry alliances are becoming an important means of assuring food safety and quality. Evidence from a recent survey of

0-8493-2217-0/01/$0.00+$.50

beef producers in Canada and the U.K. assesses producers' attitudes towards quality assurance schemes and their willingness to work interdependently within the supply chain to provide credible farm-to-retail food safety assurances.

Food Safety Systems

Introduction

This chapter highlights the different approaches to food safety in the U.K., Canada, and Australia. These countries make an interesting comparison because of their respective incentives for change and the differing approaches that government and industry have taken to ensure food safety. The contributions of economics, management, and marketing literature to understanding the incentives for change are discussed. An effective food safety system requires the participation of all parties in the supply chain, and may alter relationships between supply chain participants. Vertical industry alliances are becoming important means of assuring food safety and quality. The attitudes of beef producers in Canada and the U.K. towards quality assurance (QA) schemes are compared. Producer willingness to work interdependently within the supply chain is essential to providing credible farm-to-retail food safety guarantees.

The U.K. Food Safety System

Until recently, responsibility for food standards and safety in the U.K. was divided between several central government departments, the environmental health and trading standards departments of local authorities, and a number of other bodies. At central government level, responsibility for developing policy, promoting legislation, and implementing European Union (EU) legislation lay mainly with the Ministry of Agriculture, Fisheries and Food (MAFF) and the Department of Health (DoH), along with the Scottish, Welsh, and Northern Ireland offices.

The DoH was the lead department on issues of food hygiene, microbiological food safety and nutrition. MAFF was the lead department on food standards, chemical safety of food, food labeling, and food technology. Both departments were advised by a series of expert committees. The Meat Hygiene Service (MHS) was created as an executive agency of MAFF in April 1995 to provide a meat inspection service to licensed meat premises and enforce hygiene and welfare laws in slaughterhouses. Additional surveillance, testing, research and development, and advisory services are provided by the Veterinary Laboratories Agency, the Pesticides Safety Directorate, and the Veterinary Medicines Directorate.

The Public Health Laboratory Service, in partnership with local environmental health departments, monitored the microbiological safety of food in England and Wales. Responsibility for enforcing the majority of food legislation lies with local authorities, e.g., district and county councils. Port health authorities and environmental health departments, at district or unitary levels, enforce food hygiene legislation. Trading standards departments within county councils have responsibility for enforcing legislation of food standards and the labeling of food.

Coordination of local authority enforcement of food issues was the responsibility of the Local Authority Coordinating Body on Food and Trading Standards (LACOTS), which provided advice and guidance for enforcement authorities and advised central government on enforcement issues. Hence, a complicated, multi-layered, multi-authority system existed for the monitoring and enforcement of food safety standards.

The central piece of food legislation is the Food Safety Act (FSA) 1990. The act consolidated and updated all food legislation and implemented European legislative requirements. A key feature of the act is the emphasis on due diligence, which requires all firms in the food supply chain to have taken "all reasonable precautions and exercised all due diligence" (FSA 1990). In practical terms, this means that food retailers have instituted more extensive systems of checks on the foods that they sell.

There were three key criticisms of the U.K. regulatory structure. MAFF plays an important role in promoting the economic interests of the agriculture, fishing, and food industries, and this is particularly valuable in the international arena. However, MAFF was also responsible for protecting public health throughout the food chain. Inevitably, at times there were conflicts between concerns for food safety and the short term economic needs of some industry sectors. These conflicts were handled within MAFF, and it is not clear how they were resolved. Given the range of recent food scares, it was almost universally accepted that all aspects of food safety, (i.e., policy-making, surveillance, control, and audit) should now be separated from MAFF. This was not simply the view of consumers, public health experts, and some expert scientists, but was a widely held view in the food industry and the other regulatory organizations related to food safety (James, 1997). A new body was required which separated protection of public health and safety from promotion of industry growth, and operates openly so that decision-making is visible.

A second problem is that enforcement of food law was uneven throughout the U.K. Regulations under the FSA 1990 were enforced to varying standards between authorities. This can cause problems for nationwide food businesses despite the existence of the Home Authority Principle, by which the local authority in the area where a major food company has headquarters is responsible for issues of food law related to that company wherever it operates in the U.K. Food law enforcement competes for funding with other local authority responsibilities such as education and social services. Priorities differ throughout the country, and enforcement suffered in some areas. There was a need to raise standards of food law enforcement and ensure a consistent approach across the U.K. (James, 1997). To this end, legislation to introduce a new independent Food Standards Agency was passed by Parliament in 1999. Responsibility for food safety issues passed from MAFF to the agency when it was established in April 2000.

The Canadian Food Safety System

Responsibility for food safety in Canada is shared between the Canadian Food Inspection Agency (CFIA) and Health Canada (HC). Broad health and safety policies are the purview of HC, whose responsibilities include establishing nutritional standards, risk assessment, product labeling issues, and product recall in the event of a food safety problem. HC is mandated to assess the effectiveness of CFIA's activities relating to food safety. Founded in 1997, the CFIA is responsible for inspection and quarantine services through the Canadian Food Inspection System (CFIS), for accreditation of Hazard Analysis, Critical Control Points (HACCP) systems, and for administering the Canadian Food Inspection Act (Spriggs and Hobbs, 1999). Consolidation of responsibility for food safety under the CFIA has been criticized because the agency reports directly to the Federal Minister of Agriculture. This is perceived as a potential conflict of interest.

An important feature of the Canadian system is the respective roles of the federal, ten provincial and three territorial governments. If meat, poultry, or seafood is to be moved inter-provincially or exported, federal government regulations apply. For intra-province shipments, provincial inspection standards are usually sufficient, although some municipalities require meat to be federally inspected if it is to be sold within city limits. For a number of other food products, federal regulations also apply for intra-provincial shipments or exports, (e.g., processed fruits and vegetables, shell eggs). Currently, HACCP is voluntary, although there has been discussion of making it compulsory in all federally inspected meat processing plants. The annual report of the Auditor General of Canada released in November 1999 was sharply critical of the lack of coordination between CFIA, HC, and provincial government agencies in responding to a nationwide *Salmonella* outbreak in 1998.

The federal government planned to introduce new food safety legislation during the year 2000. An attempt to introduce the legislation in early 1999 failed due to legislative delays which prevented it from being debated before the close of the parliamentary session. Subsequent controversy over the perceived conflict of interest in having the CFIA report to the

Minister of Agriculture further delayed its introduction. If enacted, The Canadian Food Safety Inspection (CFSI) Act, would simplify Canada's food safety and inspection system by consolidating three existing federal agricultural input acts and five food-related acts into a single food act.[1] The CFSI Act reaffirms the division of responsibilities between CFIA and HC, leaving CFIA responsible for food inspection and for administering the new Act.

The Australian Food Safety System

Under the Australian constitution, state governments are responsible for the enforcement of food law. Each state has a Food Act administered by the State Health Department. The State Food Acts take precedence over all other food-related acts, such as the Meat Acts. Theoretically, state health departments have jurisdictional responsibility back to the farm. Animals are considered food when they are put on the truck to go to the abattoir; however, this can cause jurisdictional problems. For example, primary industry departments in the various states have jurisdiction for animal health, and state meat authorities have jurisdiction for slaughter and primary processing establishments. Historically, such jurisdictional questions have been settled by Memorandums of Understanding, with the Health Department only taking over responsibility when the commodity becomes a food product.

With individual states responsible for food standards, different standards emerged. In an attempt to harmonize standards nationally for domestically consumed products, ARMCANZ developed Australian Standards for primary processing establishments servicing the domestic market (e.g., fresh meat), while ANZFA has developed national food standards for further processing, distribution, and retail.[2] The Australian Quarantine Inspection Service (AQIS) monitors compliance with export standards. AQIS has been championing a HACCP/QA system for export meat plants called MSQA (Meat Safety Quality Assurance). This is a composite of CODEX-compatible HACCP and ISO-compatible quality assurance. Many domestic plants have voluntarily adopted MSQA, which eventually will be compulsory.

---

[1]These are the Feeds Act, Fertilizers Act, Seeds Act, Canadian Agricultural Products Act, Meat Inspection Act, Fish Inspection Act, and the provisions relating to food in the Food & Drugs Act and Consumer Packaging and Labelling Act.

[2]ARMCANZ is the Agriculture and Resource Management Council of Australia and New Zealand, a council of federal and state agriculture ministers. ANZFA is the Australia-New Zealand Food Authority. It began in 1991 as the NFA (National Food Agency) but was expanded in 1996 to include New Zealand treated as another state.

The Australian Standard developed by ARMCANZ for the hygienic production of meat requires the introduction of HACCP in domestic meat plants. ANZFA has developed the National Food Standards Code, which defines product standards (e.g., labeling and food composition). This code is written into each state's food act. The Food Standards Code defines product standards (i.e., related to microbiological, chemical, and physical composition), but does not define uniform hygiene standards. In theory, ANZFA could have required the Food Standards Code to apply right back to the farm. Instead it has relied on the ARMCANZ process to develop the Australian standards for the upstream supply chain. ANZFA is currently developing harmonized hygiene standards between the states. There appears to be some consensus that hygiene standards should be handled differently (less prescriptively) than product (compositional) standards. Harmonization of product standards would be an advantage, however, it has been argued that only national principles or outcomes be specified for hygiene issues. For example, hygienic beef production in north Queensland might be very different than in Victoria because production conditions are different.

The proposed new hygiene standards are based on the due diligence principle and require all food establishments to have HACCP-based food safety programs. A particular focus of debate is whether the mandatory National Food Hygiene Standard should be applied to the primary industry sector. On one side of the argument is the Health Department, which according to one observer, "wants HACCP principles to be applied absolutely throughout the food chain." On the other side is Agriculture, Fisheries and Forestry Australia (AFFA), which favors less regulatory models not requiring complex recordkeeping or registration of the farm as a food business. Conceivably, the regulatory costs could outweigh the benefits (from reduced health risks), while at the same time retailers and food processors are bringing about the desired changes in a form of market regulation as an alternative to government regulation. At the farm level, there are a number of QA schemes (e.g., Cattlecare and Flockcare) which involve HACCP-like principles to prevent chemical residue problems. Membership in such programs is voluntary, however, each farm is subject to independent audits.

Incentives for Change

Frameworks for Understanding Incentives for Change

Lessons from the Economics Literature

Market Failure occurs when externalities and imperfect information impede the provision of an acceptable level of food safety. In these cases, government intervention is justified to ensure that the socially optimal level of safety is provided. Firms may under-invest in techniques to reduce food safety hazards (or over-invest in poor quality food) when the firm's costs of producing unsafe food are less than the costs to society. As a result, too much unsafe food is produced. Negative externalities impact society in the form of higher incidences of foodborne illness, increased medical costs, lost productivity, and loss of income. Ideally, government intervention through food safety standards and inspection raises the private marginal costs for firms of producing unsafe food, so that the optimal quantity of safe food is produced. The challenge lies in measuring the relative costs and benefits of government intervention and in identifying the "socially optimal" level of food safety.

Cost Benefit Analysis can make an important contribution to the food safety debate. Society's incentive to improve food safety is affected by two types of costs. First is the cost of reducing the effects of microbiological hazards, termed mitigation costs; for example, improving production techniques, adoption of HACCP, and other risk management tools, or the use of microbiological testing and food inspection. Second is the cost of food safety problems, termed impact cost. This includes product recall, legal liability, and loss of reputation for a firm. For society this includes the negative externalities referred to above.

It is assumed that the effort expended by food firms and regulatory agencies to reduce food safety problems impacts positively on the level of food safety and on the mitigation costs of reducing microbiological contamination, and negatively on the impact costs of food safety problems. Therefore, if no effort is expended, mitigation costs are zero but impact costs are large and food safety problems occur frequently. Conversely, if food safety problems were eliminated entirely, impact costs would be zero but mitigation costs would be prohibitive (assuming this was even technically feasible). Analogous to the economic models of pollution control, this suggests that there is an optimum level of food safety for society which minimizes the combined mitigation and impact costs. Hence, the objective of a food safety strategy—both for industry and for regulators—should not be the total removal of all risks of foodborne diseases, because the costs of this strategy would outweigh the benefits. Instead, there will be an optimal level of food safety, implying that an acceptable level of food safety risk for society exists. The challenge for both industry and regulators is in identifying the optimum and designing food safety strategies to achieve it (Hobbs and Kerr, 1999).

The Economics of Information literature indicates that markets may fail to provide consumers with full information about food safety, providing a rationale for government intervention. Information asymmetry means that one party to a transaction has more information than another. For example, adverse selection arises because of hidden information prior to a transaction when a buyer does not have full information about the true quality characteristics of a good. Unable to determine good products from bad, consumers will be unwilling to pay the higher prices necessary to induce suppliers of good products to supply the market. The bad products (or lemons) chase the good products from the market (Akerlof, 1970). The analogy with the provision of safe (good) food should be clear—the market will adversely select lower quality, potentially unsafe food as a result of information asymmetry.

There are three solutions to this problem: a private market solution in which producers of good (safe) food products signal this to consumers through labeling, QA schemes, or product warranties; a public policy solution in which the industry is regulated to ensure a safe food supply; or a combination private-public solution in which private market initiatives such as labeling or QA schemes are given public accreditation. Private market solutions will be insufficient if market failure is pervasive, particularly if some firms have an incentive to act opportunistically by misrepresenting their products and riding on the safe food supply established by others. There are strong incentives for a public or private-public solution to this problem. For consumers, the incentive is to reduce information asymmetry. For honest firms who invest in brand name reputations and techniques to enhance food safety, mechanisms to prevent the appropriation of these assets by opportunistic cheater firms are important.

The *Transaction Cost Economics* literature explicitly recognizes that there are costs to carrying out business exchanges and that these costs affect the governance structures observed in an industry (Williamson, 1986). A change in the search, negotiation or monitoring and enforcement costs of an exchange can alter the incentives for forming closer vertical or horizontal alliances in an industry. Food safety is an important characteristic of food products but, due to information asymmetry, is one which consumers (and downstream food firms) cannot detect easily prior to purchase. This increases monitoring costs for downstream firms, providing an incentive for closer vertical relationships with suppliers. In some cases, regulatory penalties for failure to exercise due diligence in detecting pathogens provide food distributors and retailers with an added incentive to form closer supply chain relationships to reduce transaction costs (Hobbs, 1996).

Lessons from the Management and Marketing Literatures

While the economics literature offers insights into why markets fail to provide appropriate levels of food safety, and into the implications of market failure for society as a whole, rather less consideration has been given to the implications for individual firms that fail to deliver appropriate levels of food safety. Even less attention has been given to the opportunities for individual firms and supply chains to exploit consumer demand for increased levels of food safety and capture additional benefits from the implementation of management systems designed to deliver this. The increased awareness about food safety has created a potential source of competitive advantage. The management and marketing literature provides insights into how this competitive advantage might be captured.

From a marketing perspective, the key issue is consumer demand and the opportunities created for innovative firms that are able to provide and/or communicate "enhanced" food safety attributes most effectively. The problem that such firms face is that, unlike other quality attributes, safety is not easily measured by the buyer (Caswell, 1994). In addition, consumers may not be good risk assessors and may over-react to what they perceive as a "dread" outcome (Henson and Traill, 1993). The problem is that the consumer has to judge, not the intentions of the supplier, but his quality assurance. In response to what may be severe market disciplines, suppliers are often required to go to extreme lengths to protect the good name of their brand and exploit market opportunities. In order for such firms to capture benefits that are commensurate with their efforts, an offensive management strategy is required (Caswell and Johnson, 1991).

The search for strategies that will provide superior value to consumers has been an important contribution to the recent management literature (Christopher, 1992). The key insight of Porter's *Value Chain* concept is that competitive advantage cannot be understood by looking at a firm as a whole, or in isolation, but from the many discrete and strategically important activities that a firm performs alone and in concert with its suppliers and customers (Porter, 1985). Food safety has assumed strategic status for all food firms that plan to survive and prosper in the long term and, given the transformation that occurs from raw material to finished product, it is not something that individual firms can deliver in isolation. Moreover, the fact that it is difficult to deliver transparent food safety means that those firms that can do so are likely to capture monopoly profits, in the same way a manufacturer of an innovative new product captures monopoly profits until substitute products appear in the market. Examples of two such firms in the U.K. are presented later in this chapter.

The widespread implementation of HACCP programs, on-farm quality assurance, ISO certification, and food labeling to communicate how and when food is produced and its origins, are all examples of attempts by the food industry to demonstrate the transparent integrity of the product. The industry perceives this to be of value to consumer segments, who will be willing to pay a premium. This is also an example of a *strategic response* to a change in the market environment. Identifying strategic objectives, the means to achieve them, and the resources necessary to implement the plan of action is the realm of senior management. Managers with vision identify opportunities at an early stage and exploit them before their competitors—a skill well developed in large, multinational, branded food manufacturing companies but often missing from small and medium-sized enterprises producing commodity products for which profit margins are slim. Significantly, the international meat industry is dominated by the latter. Unsurprisingly, the one sector that has been affected most by food safety issues has been slow to recognize the opportunities that this change in the market environment presents.

The cost of compliance with food safety regulations has become a contentious issue, particularly upstream, where the incidence of compliance costs is greatest. The evidence regarding distribution of compliance costs across firms of different sizes is inconclusive (i.e., Bartel and Thomas, 1987; Food, Drink, and Agriculture Task Force, 1993; DTI, 1993). However, insufficient attention has been given to the potential management benefits from close examination of business practices, which schemes such as ISO and HACCP require. There is a small but growing literature (i.e., Mazzacco, 1996; Zaibet and Bredahl, 1997; Henson, Holt, and Northern, 1998) which supports the view that HACCP can be an effective business management tool when combined with ISO, to provide savings that far outweigh the costs of complying with food safety regulations.

Comparison of Incentive Structures and Drivers for Change

Drivers for Change in the U.K.

A decade of food safety scares, coupled with a key change in food safety legislation, has made food safety the primary concern of retailers and food shoppers. The domestic market is the primary driver of the U.K. food industry, with the multiple retailers enjoying considerable market power. Thus, it is no surprise that the two major factors behind the plethora of food safety initiatives in the U.K. are internal: the 1990 Food Safety Act and the Bovine Spongiform Encephalopathy (BSE) crisis.

The 1990 Food Safety Act (FSA)

The FSA implemented European Union Directive 89/397 on food safety and hygiene standards which were necessary for the establishment of the Single European Market in 1992. Through the FSA, the U.K. government took the opportunity to tackle rising public health concerns in the late 1980s following outbreaks of *Salmonella* in eggs and *Lysteria* in cheese. The act was intended to induce those involved in the food industry to improve their handling practices by strengthening the powers of enforcement, introducing tougher penalties, and increasing the legal responsibility for ensuring that food conforms to the provisions of the act.

The major legal change was the introduction of the "due diligence defense," which requires those engaged in food handling to be proactive in their efforts to ensure that food in their possession conforms to the provisions of the act. Previous legislation allowed buyers in the supply chain to use the so-called "warranty" defense, which only required that they prove the food was not compromised while under their control.

The 1990 FSA requires buyers to take all "reasonable steps" to ensure that the food they receive from upstream suppliers is safe. It also means that upstream firms need to demonstrate to their downstream customers that they are handling food correctly. The critical word in the definition of due diligence is "reasonable". This term is sufficiently vague that it has encouraged retailers to take extraordinary steps to ensure the safety of products by instituting stringent quality assurance programs with their suppliers, with a particular emphasis on traceability (Fearne, 1998).

The meat industry was the first to feel the impact, as retailers drew up codes of practice for their suppliers, covering all aspects of animal husbandry. The industry responded by developing generic farm assurance schemes such as Farm Assured Scotch Livestock (FASL), Farm Assured Welsh Lamb (FAWL), Farm Assured British Beef and Lamb (FABBL) and, more recently, Assured British Meat (ABM). All of these schemes cover the same critical factors—traceability, feeding, animal health and welfare, transport, and handling.

The major supermarkets now require that *all* livestock must come from suppliers who are members of a recognized farm assurance scheme. These schemes were relaunched in the 1990s under much tighter controls and independent inspections. The impetus behind the relaunch was the BSE crisis. The British food industry remains gripped by a battle to restore consumer confidence in *all* food products, not only beef, and this battle is currently forcing the pace of coordination between breeders, feeders, finishers, processors, and retailers.

Bovine Spongiform Encephalopathy (BSE)

BSE was first discovered in U.K. cattle in the mid 1980s. The number of confirmed cases rose to a peak in 1993 with over 1000 new cases each week. The BSE crisis was an important health scare in its own right because it was shown to be transmissible to humans, and because of the potentially long incubation period (up to 20 years). However, it was also important because of what it did for the public credibility of the U.K. government, supermarkets, abattoirs, and livestock producers. Of these four participants, only the supermarkets appear to have largely retained the confidence of the consuming public. They reacted swiftly and decisively to the crisis as it unfolded. The U.K. government was widely criticized for dragging its feet on the issue prior to 1996, attempting to downplay the risk to humans. When the U.K. government finally admitted to a plausible link between BSE and a new variant of Creutzfeldt Jacob Disease (CJD) in March 1996, the EU quickly announced a worldwide ban on U.K. beef and cattle exports. Since that time, a series of measures has been implemented to safeguard against BSE-infected cattle entering the human food or animal feed chains.

The BSE crisis exposed the U.K. meat industry in general, and the beef industry in particular, to intense public scrutiny. It also exposed a government that chose to gamble with public health and failed. As a result, the government has had to go overboard in an attempt to convince the general public that the problem has been resolved. Mandatory inspections by the MHS are now carried out every month by independent inspectors who assess each plant using an objective, risk-based assessment of health standards. The results of this Health Assessment Scheme (HAS) are published monthly, so those who do not meet the standards are publicly "named and shamed". However, the real significance of the BSE crisis is that it shifted the emphasis away from risk management at the retail level and the need to conform to food safety legislation, to the restoration of consumer confidence.

A fully computerized cattle passport system equipped to trace over 24 million animal movements per annum has been established. In January 1998 the industry launched, with the aid of a government grant, Assured British Meats (ABM), an impartial organization with representation from within and outside the meat industry. ABM has the sole aim of restoring consumer confidence in British meat through an industry-wide assurance scheme. This is designed to establish minimum safety standards on which retailers will not compete but will be free to "bolt on" their own QA schemes (ABM, 1998). The scheme is voluntary, but ABM hopes to attract 80% of the British meat industry into the scheme by 2001.

The BSE crisis focused the industry's attention on the importance of food safety and the devastating effects of loss in consumer confidence. Quality assurance and traceability are now top priorities for retailers. Only producers who are members of a QA scheme are eligible for the partnership arrangements which now proliferate the industry, and the race is on to develop a system for full traceability from breeder to individual meat cuts.

Drivers for Change in Canada

Several of the drivers for change in Canada are *external,* reflecting the importance of exports to the agri-food sector. Maintaining access to existing export markets, particularly the U.S., as well as obtaining access to new markets, has been a key driver. The U.S. is Canada's largest trading partner, accounting for over 50% of agri-food exports in 1998 (AAFC, 1999). For the beef sector, dependence on the U.S. export market is even higher. Major incentives for change are provided by regulatory initiatives in Canada's main export markets, which have required the adoption of equivalent measures. In the meat processing sector, this has meant that Canadian firms wishing to export to the U.S. were required to implement HACCP because it was mandatory in the U.S., even though it was still voluntary in Canada.

The Sanitary-Phyto-Sanitary and Technical Barriers to Trade Agreements of the World Trade Organization (WTO) encouraged the adoption of science-based risk assessment as a tool to manage food safety, further prodding the Canadian regulatory system to move away from its traditional organoleptic food inspection methods. Clearly, these drivers for change impact other countries, but they are particularly strong in an export-dependent country such as Canada.

An additional concern was the "national treatment" requirement of various trade agreements to which Canada is a signatory. This requires that Canada apply the same food safety standards to imports as are applied nationally to its domestic food products. The catch was that a national food safety standard did not exist—instead, an array of federal and provincial standards existed—some overlapping, some with substantial differences. There were fears that multiple Canadian standards could result in a challenge that "national treatment" was not being applied, leading to all provincial and federal standards being set to the lowest common denominator (i.e., the lowest provincial standard). Thus, international pressures are important determinants of recent changes to the Canadian regulatory system.

Governments and industry also recognized that the regulatory system should be responsive to new and emerging technologies (such as genetically modified foods). Similarly, new pathogens and improvements in microbiological science which allow the detection of these pathogens, have necessitated a complete re-thinking of food inspection. These two drivers for change are philosophically similar because they relate to intangible attributes of food which cannot be detected through sight, smell or touch, either by meat inspectors or by downstream food firms and final consumers. Assuring the safety of food or signaling the presence of intangible food attributes to consumers requires either detailed microbiological testing at each stage of the supply chain or a preventative approach to food safety. The adoption of HACCP in meat processing plants is designed to prevent food safety problems.

Important internal drivers have focused on the need to reduce duplication of responsibilities across federal government departments and to harmonize regulations nationally. Budgetary considerations and the desire to simplify the regulatory environment for firms lie at the heart of these incentives for change. Prior to the establishment of the CFIA, four federal government departments had responsibility for food safety and inspection and there were numerous methods of delivering inspection services, depending on federal or provincial jurisdictions.

Considerable criticism was levied at the pre-CFIA system, with charges that it was inefficient and, due to poor coordination and duplication of activities between the four departments, wasted scarce federal government resources during a period when eliminating the federal budget deficit was a policy priority. It was recognized that harmonizing regulations nationally would simplify the regulatory requirements facing firms, and that the failure to harmonize would likely have market access repercussions in export markets. Budgetary pressures appear to be less of a priority in the proposed new Canada Food Safety Inspection Act; instead "safe food, market access, and consumer protection" are its three stated objectives (CFIA, 1999).

Responding to the perceived needs of major markets, producer associations in the livestock industries have designed on-farm HACCP-based quality assurance systems or "Good Production Guides" for their producers. These are voluntary and of varying degrees of formality. For example, the beef industry's "Quality Starts Here" program currently does not have independent farm audits to verify that producers are following the recommended "Good Production Practices". Adoption of HACCP and other QA schemes by food processors has occurred on a piecemeal basis, depending on the industry and the requirements of major markets. Private initiatives tend to be specific to each level of the industry and are not coordinated across the producer-processor-retailer interface.

Drivers for Change in Australia

The more important drivers for change are external, though there are some significant internal drivers. The external drivers for change are related to the need for Australia to be internationally competitive in line with the critical importance of the export market to the Australian beef industry. Exports account for a large proportion of total beef production, with about 50% of production exported. This is produced in export plants which account for 80% of total industry throughput, meaning a sizable share of their output goes to the domestic market. Therefore, external drivers also have a substantial impact on domestically consumed product.

Historically, U.S. abattoir reviews have had an important influence on the Australian meat industry. This began in 1965 with the passage of the U.S. Pure Food Act which mandated a meat safety system that included international reviews of meat processing establishments by the United States Department of Agriculture (USDA). Any country exporting to the U.S. had to meet standards equal to those of the U.S. The U.S. standards became the de facto international standards because of the dominance of the U.S. market and the absence of comprehensive standards elsewhere. Australia received a jolt in early 1996 when the U.S. annual abattoir review found six export abattoirs to be unacceptable and a further eight abattoirs to be only marginally acceptable. In addition, significant deficiencies were found in the residue-testing laboratories used by the Australian meat inspection system. This stimulated the AQIS to investigate and correct what seemed to be a serious lack of control by AQIS officials, and helped foster reforms in the meat industry, including the move to HACCP/QA systems and the development of Australia's Meat Safety Enhancement Program.

One of the main international drivers for change has been the prospect of market loss due to specific problems with Australian meat. In 1981 and 1982, the species substitution scandal occurred, in which horsemeat, donkey meat, buffalo meat, and kangaroo meat were substituted for beef, which was then exported to the U.S. Only three or four small companies were involved, but it led to a Royal Commission into the meat industry and ultimately to the 1982 Export Control Act, which regulates the production and shipment of export meat from Australia. This makes clear that if an importing country's requirements are higher than the Australian standard or Codex Alimentarius, then it is necessary to comply with those importing country requirements.

In 1987 and 1994, chemical residues became a cause for concern because of a few incidents detected by USDA. These helped spawn on-farm QA programs and the National Vendor Declaration, which was launched in January 1997. The Residue Management Group was established to safeguard against chemical residue contamination in carcasses. More recently, as microbiological problems off-farm have become a more prominent issue than residue problems on-farm, this group evolved into Safemeat—a group of industry and government leaders with a mandate to provide strategic direction on meat hygiene and food safety, and to provide policy advice to government. Safemeat is viewed by some as a preferred alternative to a Single Food Agency, which, it is feared, would be overly bureaucratic and slow to respond to emerging issues.

The most important domestic driver for change is the occurrence of highly publicized food poisoning outbreaks. The most significant of these was the "Garibaldi incident" in January 1995, in which one girl died and twenty-four people were hospitalized with hemolytic uremic syndrome (HUS) contracted from mettwurst (a garlic sausage) produced by the Garibaldi Smallgoods Company. While the Garibaldi incident was a tragedy, it also has been credited with kickstarting the Australian food industry into improving food safety.

Following the Garibaldi incident, ARMCANZ decided to upgrade fresh meat inspection and hygiene standards across the country, designing harmonized domestic standards, as discussed earlier in this chapter. The Garibaldi incident has also been credited with a change in culture among smallgoods manufacturers in Australia. Manufacturers have taken the initiative to implement stringent food safety procedures ahead of mandatory regulations imposed by federal and state regulatory bodies. The Garibaldi incident, and others which followed, were clear catalysts for change. They provided the political environment in which such changes were not only possible but demanded.

The two largest supermarket retailers of beef in Australia are Woolworths and Coles. They appear to be taking lessons from the major supermarket chains in the U.K., which have developed extensive vertical alliances with their suppliers. This allows the supermarkets to push accountability back upstream by dealing with suppliers who can meet their criteria for food safety and quality. However, the supermarkets in Australia are not in a position yet to only source their supplies from beef producers who are members of a registered QA scheme. The main QA scheme for beef producers is Cattlecare. However, at the present time this only accounts for 5% of all beef cattle.

Another major domestic driver for change has been the desire to reduce the cost of food regulation. There has been a major shift in government thinking in favor of smaller government dedicated to deficit reduction. One of the manifestations of this was governments no longer willing to underwrite assistance programs for industry as they had in the past. Thus, in Australia the cost of meat inspection was gradually shifted to industry in what is called co-regulation, where the meat industry is made responsible for food safety but is backed up by government or third party audits.

Lessons from the Comparison

There are number of similarities in the development of food safety initiatives in the three countries, but there are also some important differences in the key drivers and how they have influenced the process of ensuring food safety. From the previous discussion, it would appear that the key lessons are: 1) the incidence of food safety scares; 2) differences in incentive structures; and 3) the proliferation of standards.

The incidence of food safety scares is the primary driver for change, with the U.K. arguably the most advanced of the three countries due to high profile public scares which placed the U.K. in the international headlines. Despite the pressures this created, the U.K. government failed to react quickly enough, and subsequently had to over-react with draconian measures. The Australian food industry had a taste of what a food safety scare can do, but to date Canada has not experienced a major food scare on the scale of BSE in the U.K. or Garibaldi in Australia. The key lesson here is that the U.K. suffered from being the first country confronted with a large-scale food safety crisis. The mistakes which the U.K. government made gave other developed countries a clear warning of what might happen if they did not take preventative measures. The BSE crisis in the U.K. provided the catalyst for the emergence of a global food safety agenda.

In contrast to the U.K., the Australian and Canadian food industries have a strong export orientation. Thus, while the U.K. may have been the first to feel the heat from a wave of food safety scares, other key markets, namely the U.S., followed suit and were much quicker to respond, introducing tough new standards for exporters to meet if they wanted to maintain access to the U.S. market. For both Canada and Australia, this pressure was every bit as strong as that which the U.K. food retailers brought to bear on the rest of the U.K. food industry and, given the importance of food exports to the Canadian and Australian economies, the government became involved as a facilitator rather than purely as a regulator. The key lesson is that the incentive structures were different. In the U.K., the incentives were primarily related to crisis management and the restoration of consumer confidence, while the Canadian and Australian governments were focused on risk management and the prevention of trade-threatening food safety issues.

The third aspect of food safety provision is the proliferation of standards which emerge in the absence of central government intervention. In the U.K., the 1990 FSA placed the burden of food safety on the retailers who, acting in competition rather than in concert, set about building their own unique assurance programs to protect their integrity. The proliferation of industry schemes which resulted placed additional burdens on suppliers and did not cease until ABM appeared with the task of removing food safety as a source of competitive advantage and uniting all elements of the meat supply chain under one industry-wide meat assurance scheme. In the other countries, proliferation resulted from the federal structure of regulation, with state or provincial schemes emerging in an ad-hoc fashion, until the trade loss threat focused the minds of the respective administrations to provide national support and national standards to support export markets for Australian and Canadian food products.

Why Producer Attitudes Matter

Agricultural producers are typically independent minded and value the sense of freedom of a rural lifestyle. However, the evolving requirements of a modern food safety system are a challenge to this sense of freedom. They require producers to enter into more interdependent relationships with other participants in the supply chain. Producer attitudes to the development of interdependent relationships are important in determining how far and how fast a closely coordinated food supply chain can evolve. Interdependence is becoming important in the provision of a farm-to-retail food safety system. Agricultural producers have contributed to such a system through membership in QA schemes, participation in traceability programs that include animal identification, and membership in horizontal and vertical alliances.

A recent survey examined the attitudes of beef producers in the U.K. and Canada to QA schemes and closer vertical and horizontal coordination (Spriggs et al., 1999). The objective of the survey was to see whether there were significant differences between producers in the two countries in their attitudes to closer coordination and participation in farm QA schemes. Differences were expected between Canada and the U.K. due to differences in the drivers for change and in industry structure. The incidence of food safety crises, which have been more significant in the U.K. than in Canada, is a major difference. This difference suggests that U.K. beef producers might be more likely than their Canadian counterparts to recognize the need for a QA scheme and to see themselves as part of the food supply chain. On the other hand, industry structure might lead to the opposite conclusion. Smaller production units and a greater reliance on the traditional auction system characterize the British beef industry. In Canada, the closer relationship between packer and large feedlot operators creates an opportunity for improved information flow between different stages of the supply chain, and this should enhance the ability of the supply chain to offer traceability. Hence, industry structure appears to favor the Canadian industry in developing more coordination and farm-to-retail QA guarantees.

Which of these scenarios is correct? The survey results suggest that, despite the greater fragmentation of the U.K. industry, producers have been generally more aware of the need to adopt QA schemes and trace-back initiatives. When producers were asked what management changes they had initiated in the preceding 18 months, it was clear that the changes made by U.K. respondents focused on quality assurance and traceability to a far greater extent than Canada. For example, management changes included joining a QA Scheme (44% in the U.K. and 15% in Canada); introducing mechanisms to ensure traceability (27 and 16%, respectively); joining a producer group (12 and 3%, respectively); and improving the quality of animal housing (19 and 10%, respectively).

In contrast, Canadian respondents were more likely than U.K. producers to have increased their cattle numbers, improved feed quality, and lowered production costs. The strong drivers for change in the U.K. appear to outweigh the fragmented industry structure in bringing about innovations to enhance vertical and horizontal coordination and improve quality assurance. While this historical look suggests U.K. producers place a higher priority on closer coordination than Canadian producers, this difference may be declining. When asked whether greater horizontal and vertical coordination was necessary for the future prosperity of the industry, producers in both countries agreed strongly, and with about the same degree of conviction.

The respondents in the two countries had very different ideas about what should be the objective of a QA scheme. Respondents were asked to choose one phrase from among several which best characterized the purpose of a QA scheme. The top two choices were to convince consumers that beef is *safe* and to ensure that only the highest *quality* beef enters the food chain. Interestingly, Canadian respondents picked quality ahead of safety (53 to 22%), while in the U.K. the order was reversed (22 to 50%). These results reflect significant differences in the nature of QA schemes developed in these two countries. In the U.K. where safety is of paramount importance, third party audits of on-farm beef QA programs are mandatory. In Canada, where food safety is not such a highly charged issue, third party audits are not mandatory. This difference is understandable. If the focus is on food safety, the benefits are largely public, so strong public accountability would be considered a high priority. However, if the focus is on improving production methods to create a higher quality product, the benefits are largely private, and strong public accountability would be less important.

This difference in perceived benefits is borne out in further results. Respondents were asked what they thought were the main benefits of a QA scheme. The results for current members are presented in Table 3.1. Focusing on current members, it highlights why producers may have joined the scheme. The most popular choice in the U.K. was "to improve consumer confidence," while in Canada it was "to provide information to improve production".

Respondents were also asked what would be the main costs (problems) of a QA scheme. Table 3.2 shows the responses of respondents who are *not* currently members of a QA scheme. Focusing on non-members, it highlights why producers may not have joined the scheme.

Table 3.1 Benefits of Belonging to a QA Scheme (Current Members)

Perceived Benefit	*Canada*	*UK*
More secure markets	2.2	2.2
Improved consumer confidence	2.7	2.6
Information to improve production	3.4	1.9
Compliance with food regulations	2.7	2.4
Premium above normal market prices	2.8	2.0
Stronger links with the trade	2.9	2.2

(Average score where 1 = not significant, 5 = highly significant)

Table 3.2 Costs of Belonging to a QA Scheme (Non-Members)

Perceived Benefit	*Canada*	*UK*
Inconvenience of farm inspections	2.4	2.8
Training self/staff to meet standards	2.5	2.4
Increased capital investment	2.8	2.9
Reduced independence	3.3	3.3

(Average score where 1 = not significant, 5 = highly significant)

There is broad agreement by non-members in both countries regarding the most significant costs of QA scheme membership, with "reduced independence" the most popular choice. This supports the assertion that agricultural producers value their independence and that overcoming this desire for independence is an important challenge for those interested in encouraging greater coordination of the supply chain.

The difference in U.K. and Canadian producer attitudes to the purpose of a QA scheme may create competitive difficulties for Canada. In the wake of a number of food safety scares around the world, there appears to be a growing international demand for food safety systems to be publicly accountable. If the Canadian beef industry is to remain internationally competitive, it must address the question, how can producer attitudes be modified so producers will accept a QA scheme with strong public accountability? Producer attitudes *do* matter in the development of a food safety system. Their attitude to working interdependently with other participants in the supply chain is essential for providing credible farm-to-retail food safety guarantees.

The Role for Alliances

Vertical Alliances in the U.K. Food System

Some players in the U.K. beef industry have recognized that a change in emphasis is required, towards the benefits which can come directly from addressing consumer requirements for food safety and indirectly from the systems which have been put in place to deliver safe food. This change in emphasis is manifested in the growing importance of supply chain partnerships, from retailers, farmers, and breeders to feeders and other input suppliers.

The major retailing multiples and the largest abattoirs have yet to find an effective system for tracing products from the breeder through the cutting plant to the retailer on a commercial scale. This has provided the smaller players with an opportunity to gain a competitive advantage.

One such player is Tracesafe, a farmer-owned company in S.W. England, which has operated a unique cattle traceability and QA system since January 1996.

The Tracesafe Cattle Management System is the first of its kind in the U.K. to receive the internationally recognized ISO 9002 quality assurance accreditation, covering parent selection and all stages of rearing and production through to receipt of the carcass by the processor or butcher. Systems are in place to allow the history of individual meat cuts to be traced back to the animal of origin. Tracesafe beef is targeted to specialist retail outlets and high quality restaurants, where consumers are willing to pay a premium for the assurance of guaranteed traceability. There are currently 130 members, including breeders, breeder/finishers, and finishers. Calves of any breed are supplied from units which comply with MAFF welfare standards. All animals are reared on natural feed. On-farm feed mixing is encouraged, and all grain is supplied from a network of mills contracted to provide specially prepared rations (free from hormones, growth promoters, or fishmeal) into breeding and rearing units, where independent auditing is carried out under the ISO 9002 accreditation requirements. Tracesafe has a unique computer-controlled birth card system which records the dam and sire of every calf, and allows the animal to be monitored through every stage of the rearing and meat-processing chain. Complete details of an animal's life, including parentage, medication, feeding, and any movements are fully documented. The BSE and tuberculosis (TB) risk is minimized with cattle supplied from parentage that can have either one, two, or three generations free of BSE and TB, as required. The brand name, Tracesafe, serves as the quality assurance stamp to be used on all accredited carcasses. This helps prevent fraud and acts as an endorsement of traceability.

At the other end of the supply chain, Marks and Spencer (M&S) are probably the closest to having a system similar to that developed by Tracesafe. Unlike other major food retailers who work mostly with abattoirs, M&S has a direct link with their farmer suppliers. The M&S' select beef scheme differs from the Tracesafe initiative because it was originally designed to deliver traceability for the purpose of improving product quality.

M&S' select beef scheme claims to deliver consistently high quality meat backed by frequent taste panel tests which, using a detailed producer database, can be related directly back to the individual farm. Every producer who applies for approval submits information covering housing, breeds, feed use, and stockmanship. M&S claims its code of practice is superior to generic assurance schemes such as FABBL. They use the privatized Agricultural Development and Advisory Service (ADAS) to carry out random inspections, but initial inspections of applicants are carried out by the farm assurance officer of the local abattoir approved to process meat for the scheme. Results of taste panel tests are used to compare beef produced under different regimes, enabling technical staff to recommend changes to a ration or husbandry to further ensure a consistent eating quality. Every six months, producers are asked to complete a feed declaration, and details are entered on the database. Any changes are then highlighted. When buying feed, producers must have a breakdown of all ingredients to show that only approved ingredients are used. These two examples demonstrate an important change in the nature of contractual relationships in the British meat industry, away from adversarial spot trading to tightly organized and highly integrated strategic alliances, which have responded swiftly and effectively to the demands for improved safety.

Vertical Alliances in the Australian Food System

Vertical partnerships are emerging in the Australian beef industry and are being led by the major supermarket chains, Woolworths and Coles. In the case of meat, this is happening in part because the Australian government has been encouraging co-regulation whereby the meat industry is required to take direct responsibility for the production of safe food. In addition, the large supermarket chains are following the lead shown by their U.K. counterparts. For example, Woolworths has developed a Vendor Quality Management Standard (WVQMS) for its suppliers. Suppliers who agree to the program implement a HACCP plan, which is subject to a verification audit by food safety auditors appointed by the retailer plus an independent third party audit.

Meat and Livestock Australia (MLA) has encouraged vertical partnership initiatives through its Marketlink Program. The nature and extent of retailer-led beef partnerships varies depending on the location. In northern Australia (Queensland), it is relatively more difficult to get consistent product. Hence, there is a stronger incentive for the major retailers to form vertical alliances with feedlots and processors. By contrast, in the south (Victoria), the beef cattle are grass-fed, production is highly seasonal, and the animals are typically marketed in small lots through saleyards. There is less scope for vertical partnerships with producers.

Another initiative developed by MLA to encourage consistent quality meat has been Meat Standards Australia (MSA). This is a gate-to-plate grading and trading system based on Palatability Assured Critical Control Points (PACCP) rather than HACCP, which uses control systems throughout the supply chain that focus on eating quality. Using extensive consumer taste tests, a set of objective criteria has been determined which are highly correlated with eating quality. These criteria relate to livestock production and processing characteristics. The carcass is graded according to these objective criteria and an eating quality index is determined for each cut of meat. The eating quality index is used to determine the grade placed on the cut of meat for retail sale. A pilot study conducted by MLA in Brisbane in 1998 revealed consumer willingness to pay for consistent quality. This program was subsequently implemented nationally.

From the beef producer's perspective, Marketlink is exclusive, while MSA is inclusive. Marketlink involves vertical alliances between particular parties to an agreement, while MSA is open to any producer who meets the production criteria. The MLA is currently more actively promoting the latter as the option for achieving closer vertical coordination in the beef industry. However, the two approaches are not mutually exclusive. MSA could provide a basic guarantee of eating quality, but the retailers may wish to add their own specifications to differentiate their product. This could be accomplished through the type of vertical partnerships envisaged by Marketlink.

The Changing Role of Government

Maintenance of an effective and credible food safety regulatory system remains a critical role for governments. The challenge lies in designing a system which assures consumers of a safe food supply while avoiding draconian measures that hamper industry competitiveness with little marginal benefit in improved safety. Government responses to food safety crises are critical in maintaining consumer confidence in the food sector. The BSE crisis provides a vivid example of the consequences of a failure in consumer confidence. However, food safety is by no means a government-only responsibility. All players in the food supply chain, from input providers to retailers (and, arguably, consumers) have a responsibility to ensure that food is safe. There are strong market, supply chain, and regulatory incentives for food firms to do this. Fundamentally, ensuring a safe food supply and maintaining consumer confidence requires improved information flow through closer supply chain relationships. An evolving role for government may be in the accreditation of private sector supply chain initiatives to enhance food safety and assure quality.

Conclusions and Future Research Needs

This chapter has outlined the differences in institutional arrangements for ensuring food safety in the U.K., Canada, and Australia. National harmonization of standards, increased private sector accountability, and tighter regulatory control are features of all three systems. The industry response has been swifter and more decisive in the U.K. than in Canada or Australia, and has been largely driven by the retail sector. Paradoxically, given that retailers are the final point of contact with consumers, they have not featured prominently in Canadian industry QA initiatives. Until retailers become involved, these initiatives cannot truly be regarded as complete "gate-to-plate" supply chain partnerships.

It is not clear which institutional environment will be the most effective and efficient means of delivering safe food. Further research is needed to determine the relative effectiveness of different national systems. Consumer confidence is an elusive concept, yet it is critical to the sustainability and competitiveness of a food industry. Further research is warranted into how consumer confidence is created, how consumer attitudes differ across national environments, and the extent of the potential price premium for assured safe food or whether this is simply a baseline market expectation. In the international trade arena, food safety and consumer preferences are critical and controversial issues, and therefore will be a likely source of trade disputes. In this context, it is essential that we have a clearer understanding of consumer preferences and the consequences of diverging national policies to ensure food safety.

References

Agriculture and Agri-Food Canada (AAFC), *A Portrait of the Canadian Agri-Food System,* AAFC http://aceis.agr.ca/policy/epad/english/pubs/chrtbook/chrtindx.htm, 1999.

Akerlof, G.A., The market for "lemons": qualitative uncertainty and the market mechanism, *Quarterly Journal of Economics,* 84, 488, 1970.

Assured British Meats (ABM), *Meat Industry Sets New Standards with Consumer Assurance Body,* press release, ABM, Milton Keynes, 1998.

Bartel, A. and Thomas, L., Predation through regulations: the wage and profit effects of the occupational safety and health administration and the environmental protection agency, *Journal of Law and Economics,* 30, 239–264, 1987.

Canadian Food Inspection Agency (CFIA), "Backgrounder: Development of the Canada Food Safety and Inspection Bill", http://www.cfia-acia.agr.ca/english/reg/legren/backaidee.shtml, CFIA, Ottawa, Canada, 1999.

Caswell, J. The policy environment for food safety and nutrition: regulating quality and quality signalling, in Padberg, D., Ed., *Re-engineering Marketing Policies for Food and Agriculture,* Texas A&M University, 57–76, 1994.

Caswell, J. and Johnson, G., Firm strategic response to food safety and nutrition regulations, in Caswell, J., Ed., *Economics of Food Safety,* Elsevier, New York, 273–297, 1991.

Christopher, M., *Logistics and Supply Chain Management,* Pitman, 1992.

Department of Trade and Industry (DTI), *Review of the Implementation and Enforcement of EC Law in the UK,* HMSO, London, 1993.

Fearne, A., The evolution of partnerships in the meat supply chain: insights from the British beef industry, *Supply Chain Management: An International Journal,* 3(4), 214–231, 1998.

Food, Drink, and Agriculture Task Force, *Proposals for Deregulation,* Department of Trade and Industry, London, 1993.

Henson, S, Holt, G., and Northern, J., *Costs and Benefits of Implementing HACCP in the UK Dairy Processing Industry,* paper presented at Economics of HACCP: New Studies of Costs and Benefits conference, Washington, June 16–18, 1998.

Henson, S. and Traill, B., The demand for food safety: market imperfections and the role of government, *Food Policy,* 18(2), 152–162, 1993.

Hobbs, J.E., A transaction cost analysis of quality, traceability and animal welfare issues in UK beef retailing, *British Food Journal,* 98(6), 16, 1996.

Hobbs, J.E. and Kerr, W.A., Costs/benefits of microbial origin, *Encyclopedia of Food Microbiology,* in Robinson, R. and Patel, P., Eds., Academic Press, London, 480–486, 1999.

James, P., *Food Standards Agency Report,* HMSO, London, April, 1997.

Mazzacco, M., HACCP as a business management tool, *American Journal of Agricultural Economics,* 78, 770–774, 1996.

Porter, M., *Competitive Advantage,* The Free Press, 1985.

Spriggs, J. and Hobbs, J.E., *Competitiveness of Canada's Beef Food Safety System: Comparison with the United Kingdom,* Agriculture and Agri-Food Canada, Ottawa, May 1999.

Spriggs, J., Hobbs J.E., and Fearne, A., *Beef Producer Attitudes to Farm Assurance Schemes in Canada and the UK,* presented to the International Food and Agribusiness Management Association Annual Conference, Italy, June 1999.

Williamson, O.E., *Economic Organization: Firms, Markets and Policy Control,* Harvester Wheatsheaf, Hemel Hempstead, UK, 1986.

Zaibet, L. and Bredahl, M., Gains from ISO certification in the UK meat sector, *Agribusiness,* 13(4), 375–384, 1997.

Chapter 4

Quantifying Phytosanitary Barriers to Trade

Hugh Bigsby and Carolyn Whyte

The Sanitary and Phytosanitary Agreement (SPS) of the General Agreement on Tariffs and Trade (GATT) is an effort to reduce the technical barriers to trade created by phytosanitary regulations. A key feature of the SPS agreement is the role of risk assessment and risk management in determining appropriate quarantine actions that provide an acceptable level of risk to the importer and can be justified on technical and trade terms. A major problem to date has been quantifying the effects of phytosanitary regulations in a way that permits objective comparisons. This chapter presents a model for quantifying quarantine-related trade barriers. The model combines the two basic components of pest risk assessment, probability of establishment and economic effects, into a management framework and an objective measure and provides a systematic basis for defining and measuring acceptable risk and justifying quarantine actions relative to acceptable risk.

Introduction

One of the outcomes of the Uruguay Round of the General Agreement on Tariffs and Trade (GATT) was the provision for reductions in a range of

0-8493-2217-0/01/$0.00+$.50

trade barriers. Barriers such as tariffs, export subsidies, embargoes, import bans, quotas, supply management regimes, domestic price supports, and licensing and exchange controls were dealt with by converting them into tariff-equivalent levels of protection through a system of "tariffication." The key success of this approach was that different quantifiable trade barriers could then be compared, reduced, or negotiated in a common framework.

What remained to be resolved was a range of trade barriers that were largely non-quantifiable in terms of tariff-equivalent levels of protection. These barriers, termed Technical Barriers to Trade (TBT), included rules and standards directed at health, safety, or the environment. A key feature of TBTs which differentiates them from the quantifiable trade barriers is that they are not specifically targeted at trade or production issues. Under GATT rules, countries are generally allowed to adopt health, safety, or environmental policies which take precedence over other rules. The caveat to this, however, is that these policies are only allowed as long as the purpose of the policy or standard is to meet a legitimate domestic objective, and as long as domestic and foreign producers are treated in the same manner.

Among the most prevalent of the TBTs are requirements that deal with concerns about human, animal, and plant health (Hillman, 1978, 1991). Concern has been raised that with the reduction in quantifiable barriers to trade, countries will turn to TBTs as a way of blocking imports rather than just meeting legitimate sanitary and phytosanitary concerns (Ndayisenga and Kinsey 1994). This concern has led to major efforts internationally to ensure that sanitary and phytosanitary measures do not evolve as major trade barriers.

Under GATT and its successor, the World Trade Organization (WTO), technical barriers to trade related to animal and plant health issues are dealt with under the Sanitary and Phytosanitary (SPS) Agreement. Under the umbrella of the SPS Agreement, the International Plant Protection Convention (IPPC) has produced standards for determining the Appropriate Level of Protection (ALP), or justified quarantine measures for plants (Food and Agriculture Organization [FAO], 1996), and the International Office of Epizootics (OIE) is doing the same for animals. The major problem faced by the IPPC and the OIE is the lack of a system that can convert the diverse technical or scientific barriers related to plant and animal health into a common framework of ALP, which would allow for the comparison of quarantine measures within a trade or economic forum. A common theme of the activity of the IPPC and the OIE is a need to develop systems that will measure ALP and show whether phytosanitary or animal health standards are being imposed in a way that is consistent with both internal and external standards.

Another of the key changes under the Uruguay Round of GATT has been a focus on risk assessment and management, with an overall objective of minimizing negative trade impacts (Papasolomontos, 1993). This is a considerable departure from past practice in the quarantine area, where historically SPS has been an activity of scientists with a focus on assessment of probability of occurrence as the key criteria for applying trade barriers (Smith, 1993; Patterson, 1990). This is an objective but one-sided application of standards in a trading environment. The changes under the Uruguay Round mean that risk assessment now requires consideration of economic consequences as well as probability of occurrence. In addition, risk management now requires the consideration of trade-offs in probability of establishment and economic consequences, and in the context of choosing the least trade-distorting path.

As a result of these changes, New Zealand's Ministry of Agriculture and Forestry (MAF) began an interdisciplinary research program that would combine economic analysis with traditional scientific approaches to determine phytosanitary (plant-related) risk (Bigsby and Whyte, 1998; Bigsby, 1996a, 1995a,b, 1994b; Bigsby and Crequer, 1995; Greer and Bigsby, 1995; Greer et al. 1995) and assisted in work for the IPPC (Bigsby, 1994a). The key features of this research program were expansion of the expertise involved in quarantine risk assessment and the development of a mechanism that could link both economic and scientific components of a risk assessment in a unified measure of risk. In the case of phytosanitary risk, the interdisciplinary team requires a range of expertise that could cover the entire spectrum of trade in plant products.

- Analysis of existing pests and diseases in the exporting country, and their life cycles and environmental requirements to determine whether anything could potentially survive in the importing country.
- Assessment of trade patterns, shipment methods, and storage of goods to determine whether there is potential for a pathway to allow introduction of a pest or disease.
- Assessment of the potential physical effect of exotic pests and diseases on new hosts in the importing country, should something be introduced.
- Analysis of the potential economic consequences of introduced pests and diseases.

This chapter presents a risk analysis system, the Iso-Risk Framework, that combines interdisciplinary inputs into a single analytical system, providing a quantifiable measure of the level of protection associated with a quarantine measure.

Iso-Risk Framework

A key factor in assessing levels of protection is development of a methodology that uses both economic effects and probability of introduction to manage risk (FAO, 1996). Although the FAO's draft standards do not specify how to combine economic effects and probability of introduction, the implication is that they should be considered together to measure Pest Risk.

A common way for these two factors to be combined is to calculate pest risk as,

$$\text{Pest Risk} = \text{Economic Effect} \times \text{Probability of Introduction}$$

Use of both the probability and consequences of a particular event to express risk appears in many areas of risk analysis (Kaplan and Garrick, 1981; Cohrssen and Covello, 1989; Miller et al., 1993; Ministry of Health, 1996). The framework discussed here follows this approach and discussions during the development of the draft Pest Risk Analysis Standards by the IPPC working group (Orr, 1995). This framework has been further developed in New Zealand (Bigsby, 1996a; Bigsby and Crequer, 1996).

Calculated this way, pest risk represents the expected economic effect of pest introduction during the time period for which the probability of introduction has been assessed. Management options considered by a quarantine authority using this definition would change pest risk towards some benchmark or acceptable level (equivalent to "acceptable level of protection", ALP) by altering the probability of introduction or the economic consequences of establishment. A critical component is the establishment of a benchmark level of acceptable pest risk so that subsequent management strategies can be systematically evaluated against the benchmark.

Pest Versus Commodity Risk Assessment

Many quarantine risk assessments focus on the risk associated with a particular pest. However, trade restrictions and most pre-entry quarantine measures are directed at entire commodities rather than particular pests. A "commodity" here refers to a specific product and country/pathway combination. In particular, commodities with more types of pests will represent a greater risk per unit than commodities with fewer types of pests. A purely pest-based analytical approach, while useful for some types of analyses, such as categorizing pests into quarantine and non-quarantine, may not give a measure of the overall risk associated with a commodity.

Commodity-based risk assessments, such as those produced by the U.S. Department of Agriculture (USDA, 1995), rely on assessments of each pest associated with a commodity. Similarly, the appropriate level of protection

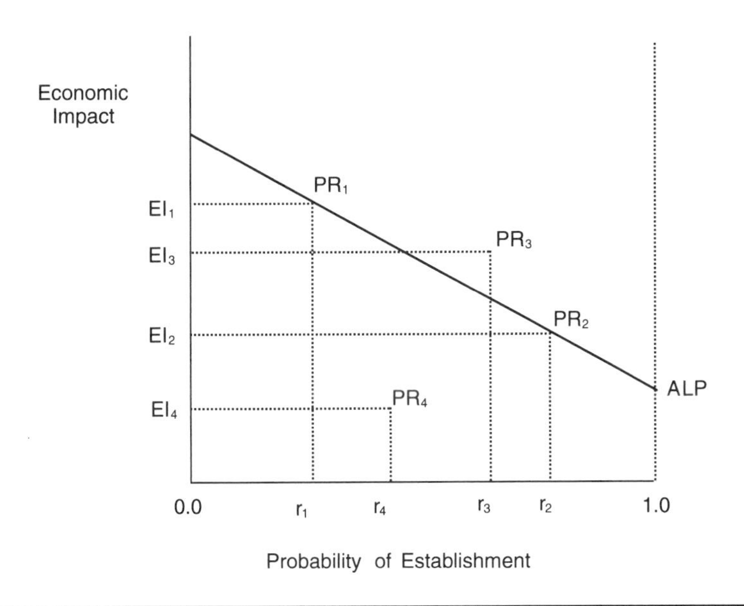

Figure 4.1 Iso-Risk Framework

can be defined for a commodity by considering the appropriate levels of protection for each individual pest of the commodity.

Pest Evaluation

The basic concept of the iso-risk framework is illustrated in Figure 4.1. Pest 1, with an economic impact of EI_i and a probability of establishment of r_1, has a pest risk of PR_1[1], where

$$PR_1 = EI_1 \times r_1$$

[1]Pest Risk is depicted in Figure 4.1 as a point estimate or single value. This is done for purposes of illustration in developing the general methodology in this chapter. In practice, there would be a problem in providing only a point estimate because it gives no quantitative picture of the uncertainty surrounding either the probability of establishment or economic impact values used in the pest risk estimate. This means that there is no information on whether a particular estimate represents the most likely value, or one of a host of equally likely values over a wide range (i.e., Cohrssen and Covello 1989). Since pest risk is actually based on a probability distribution for both risk of introduction and economic impact, rather than being a point estimate, a plot of pest risk would be an area. Given a distribution of outcomes, a decision maker would be in a position to make a better-informed assessment of the appropriate management actions for a particular pest than with only a point estimate.

Pest 2 has an economic impact of EI_2 and a probability of introduction of r_2. Different pests, having different potential economic consequences and probability of introduction, may still share the same value of pest risk.

In Figure 4.1, PR_1 and PR_2 lie on a line along which all combinations of ($EI_i \times r_i$) have the same value (hence, the iso-risk line). Note that the iso-risk line is straight only when both the x and y axes are plotted with logarithmic scales.

An important requirement for determining entry conditions is a predetermined benchmark level of pest risk, or ALP. In Figure 4.1, there can be any number of iso-risk lines representing different levels of pest risk, with higher iso-risk lines indicating higher pest risk. The iso-risk lines allow pests to be ranked with respect to each other and to a particular acceptable level of pest risk. This provides the basis for determining suitable entry conditions (i.e., those resulting in a value of pest risk after risk management that does not exceed the ALP, with a reasonable level of confidence). Since all points on an iso-risk line have the same expected value,

> *The ALP represents the highest iso-risk line that will be accepted by a quarantine authority.*

Evaluating individual pests against the ALP is then straightforward. If the pest risk of a particular pest is greater than the ALP, actions should be taken to reduce pest risk to the ALP. For example, if the iso-risk line in Figure 4.1 has been determined to be the ALP, a pest with a pest risk of PR_3 would be subject to actions to reduce the risk to acceptable levels. The pest corresponding to PR_4 falls within acceptable limits and requires no additional quarantine actions.

Commodity-Based Risk Assessment

The pest risk of a commodity (PRC) can be considered as the cumulative expected value of all the associated pests for that commodity. If PRC is the expected value of pest risk for a commodity, then,

$$PRC = \sum_{i=1}^{n} (R_i \times EI_i)$$

where R_i is the probability of establishment of pest i, EI_i is the economic impact of pest i, and n is the number of pests associated with the commodity. Since PRC is the sum of a number of individual pest risks, it can take any value from 0 to ∞, as is shown in Figure 4.2.

Using this approach, a quarantine authority could consider commodities having similar values of PRC, regardless of the number or type of pests involved, with the same level of concern. A benchmark ALP can also be

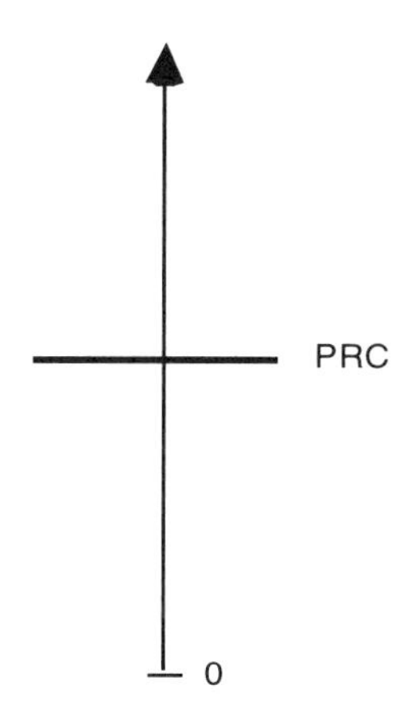

Figure 4.2 Commodity Pest Risk (PRC)

defined for commodities as follows:

> *The ALP is the highest value of commodity pest risk that will be accepted by a quarantine authority.*

In Figure 4.2, the ALP would represent a cut-off point on the axis. Appropriate entry conditions would ensure that the commodity risk does not exceed the ALP with a reasonable level of confidence.

Implementing Iso-Risk as a Trade Tool

Appropriate quarantine actions will be those that ensure the pest or commodity risk does not exceed the ALP, with a reasonable level of confidence. Either the probability of introduction or the economic impact could, in theory, be modified by a particular quarantine action. Pre-entry measures, such as area freedom, quality systems used during production and after harvest, and disinfestation treatments, are aimed at reducing the probability of introduction. Quarantine actions that reduce the expected economic impacts of pests tend to occur within the importing country, rather than before entry, and include surveillance systems for early detection and official eradication or control programs. These actions are designed to eliminate or at least reduce the spread of the initial outbreak, thus limiting the economic consequences.

Pre-entry quarantine actions are directed at commodities, therefore any single measure will potentially impact more than one pest associated with the commodity. In-country measures, however, are directed at specific pests. Thus, evaluation of quarantine strategies may require consideration of appropriate levels for both individual pest risk and commodity risk.

Modifying the Probability of Establishment

Using the iso-risk model, a key risk management tool that can be applied is modification of the probability of establishment. The probability that an exotic pest establishes in a new country is dependent on the number of units of the commodity imported, the proportion of those units infested with the particular pest, the number of individuals per infested commodity, and the suitability of the new environment for the pest. The first two factors determine the number of potential sources for establishment, while the latter two express the likelihood that any particular source results in an establishment.

As presented in Baker et al. (1993), the probability of exotic pest establishment (r) can be calculated as,

$$r = 1 - (1 - p\,\Phi)^N \quad (4.1)$$

Where p is the proportion of commodity units infested, Φ is the probability that a single infested unit leads to an establishment, and N is the number of units imported. To reduce r, then, one must reduce either p, Φ, and/or N.

The calculation of Φ depends on factors such as the number of pests present on an individual unit (μ), the probability that an individual pest survives to reproduce in the new country (including survival of transit, natural mortality, predation, and parasitism), collectively called ϕ, and the probability that the pest finds suitable abiotic conditions for survival and establishment (including suitable climate, hosts and soil conditions), collectively called Ψ. Methods for calculating Φ are given in Whyte et al. (1996). To reduce Φ, thereby ensuring that the individual pest risk (PR) does not exceed the maximum acceptable level (ALP), the values of μ, ϕ, and/or Ψ must be reduced. This results in five factors that quarantine authorities may manipulate in order to ensure that pest risk, and ultimately commodity risk, does not exceed the acceptable level.

- p proportion of units in a commodity infested with a particular pest
- N number of units in a commodity imported in a particular time period
- μ average number of individuals of a particular pest per infested commodity
- Φ probability that a particular individual survives to reproduce
- Ψ probability that conditions (climate, hosts) are suitable for pest survival

The way specific quarantine measures are used to reduce pest risk to meet ALP can be illustrated in an example. Consider a hypothetical pest-commodity combination using the following assumptions.

Appropriate Level of Protection (ALP)	$100 per annum
Economic impact of the pest (EI)	$1,000,000 per annum
Amount of commodity imported annually (N)	500,000 units
Average number of individuals per infested unit (μ)	10
Proportion of individuals surviving to reproduce (ϕ)	0.3
Suitability of conditions for the pest (Ψ)	0.5

We assume for simplicity that the consequences of pest establishment, determined as per Greer and Bigsby (1995), will not be reduced by a surveillance network or a pre-planned outbreak response system. Therefore, if ALP = r × EI, the probability of establishment, r, must be reduced to no more than (100/1,000,000) or 0.0001. Given this information, the quarantine measures required to reduce the probability of establishment to 0.0001 can be determined.

The first step is to calculate Φ and r with the available information. We will assume that the pest is biparental. Using the method[2] of Whyte et al. (1996) to calculate Φ results in a value of $\Phi = 0.3$. Fresh produce and many other plant products are inspected on arrival in New Zealand using a 600-unit sample. If the pest can be easily detected by inspection and no pests are found in the sample, then the upper 95% confidence limit for the infestation level, p, can be calculated as $1 - \sqrt[600]{(1 - 0.95)}$ (Couey and Chew, 1986), or 0.005. This gives a value for r, calculated using Equation 4.1, of 1. Clearly, inspection on arrival is insufficient as a quarantine measure on

[2]A general calculation for biparental organisms, including tephritid fruit flies, is shown below:

$$\Phi = \Psi \times \sum_{x=2}^{\infty} f(\mu\phi)x(1 - 0.5^{x-1}) \tag{4.2}$$

where $f(\mu\phi)$ is an appropriate distribution function describing the expected number of survivors per infested unit. If the number of survivors per infested unit is assumed to follow a Poisson distribution,

$$\Phi = \Psi \times (1 + e^{-\mu\phi} - 2\,e^{-\mu\phi/2})$$

(Whyte et al. 1996)

its own, and additional quarantine actions are justified. There are several different ways to determine appropriate entry conditions for this pest.

Reduce the Infestation Level

The infestation level required to reduce the probability of establishment to the acceptable level can be determined by rearranging Equation 4.1 to solve for p (Baker et al. 1993).

$$p = \frac{1 - \sqrt[N]{1 - r}}{\Phi} \tag{4.3}$$

The value of p found is the maximum acceptable infestation level sufficient to ensure that the acceptable level of risk is not exceeded (called the Maximum Pest Limit by Baker et al., 1993). Setting r to 0.0001, using $\Phi = 0.3$ and $N = 500{,}000$ gives a value of $p = 6.6 \times 10^{-10}$.

Data may be available from the endpoint inspections associated with quality production systems or from pre-export inspections of the commodity to determine the actual infestation level. The number of units (n) to inspect to be 95% confident that the infestation level of a commodity is no more than p can be calculated as

$$n = \frac{\log(1 - 0.95)}{\log(1 - p)} \tag{4.4}$$

(Couey & Chew 1986)

Therefore, to be 95% confident that the infestation level was no more that 6.6×10^{-10} would require inspection of a total of 4.5 billion commodities, with none found to be infested. Clearly this is impractical in the short term, although it could possibly be accomplished over a number of years, and so other factors must be reduced to reduce r to 0.0001.

Reduce Survival with a Post-Harvest Treatment

Post-harvest treatments reduce the survival of individuals, and so affect the value of Φ. To determine the required efficacy of a treatment, Equation 4.3 can be rearranged to solve for Φ, as below.

$$\Phi = \frac{1 - \sqrt[N]{1 - r}}{p} \tag{4.5}$$

The target value of r is 0.0001. If inspection can detect the pest on infested items, $p = 0.005$ and $\Phi = 4 \times 10^{-8}$. If inspection can not detect the

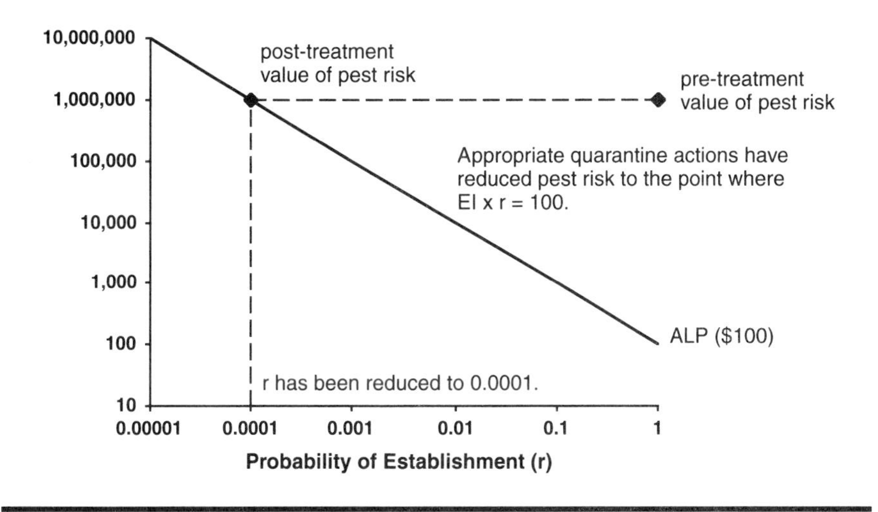

Figure 4.3 Effect of Entry Conditions on Pest Risk

pest with a high degree of reliability, $p = 1$ and Φ must be 2×10^{-10}. Now we can determine the value of ϕ necessary to produce these values of Φ. Note that Equation 4.2 cannot be rearranged to provide an algebraic solution for ϕ, and the calculations must be done on a computer. If inspection can detect the pest on infested items, the value of ϕ necessary to give $\Phi = 4 \times 10^{-8}$ in Equation 4.2 is 5.7×10^{-5}. Without inspection, Φ must be 2×10^{-10}, and $\phi = 4 \times 10^{-6}$. The value of ϕ is calculated as the product of several survival probabilities (assumed to be independent, Whyte et al., 1996). The new value of ϕ is calculated as the product of post-harvest treatment efficacy (proportion surviving treatment) and the pretreatment value of ϕ.

In this example, the value of ϕ without a post-harvest treatment is 0.3. Therefore, the post-harvest treatment must have an efficacy of $5.7 \times 10^{-5} \div 0.3$ (with inspection) or $4 \times 10^{-6} \div 0.3$ (without inspection). In the first case, the efficacy is 0.00019 (equivalent to a treatment strength of probit 8.6)[3], while in the second, the efficacy is 1.33×10^{-5}, equivalent to a treatment of probit 9.2. If the pest can be detected visually, the 600-unit inspection on arrival acts to reduce the required treatment from probit 9.2 to

[3]Quarantine treatment data have historically been evaluated using probit analysis, in which the normal distribution is used to transform the mortality variable to a linear relationship with the dose. A mortality of 50% corresponds to a probit value of 5, while 99.9968% mortality corresponds to probit 9. This value has been used as the standard for quarantine treatments since Baker (1939) advocated it as a suitable treatment mortality for tephritid fruit flies.

probit 8.6. Figure 4.3 shows the pre- and post-treatment values of pest risk for this pest. The value of r is now 0.0001 and EI $\times$ r = ALP, so the appropriate entry conditions are, in this case, either a post-harvest treatment of efficacy 0.00019 plus inspection of a 600-unit sample on arrival, or a post-harvest treatment of efficacy 1.33×10^{-5}.

Suppose data were available showing that over a 3-year period, 100,000 units of the commodity had been inspected prior to export and found to be uninfested. The observed infestation level is 0, and the upper 95% confidence limit (p) is calculated as

$$p = 1 - \sqrt[N]{1 - 0.95} \tag{4.6}$$

or 3×10^{-5}. The value of r, from Equation 4.1, is now 0.989. Use of the pre-export clearance data seems to have had little effect. However, even this small adjustment in r has an effect on the required post-harvest treatment efficacy. Using $p = 3 \times 10^{-5}$ in Equation 4.5 and solving for Φ results in $\Phi = 6.7 \times 10^{-6}$. The value of ϕ necessary to make $\Phi = 6.7 \times 10^{-6}$ in Equation 4.2 is 0.00073. The required treatment efficacy is therefore $0.00073 \div 0.3$ or 0.0024, equivalent to probit 7.8. Thus, although adjusting p appeared to make little difference in terms of r, the required treatment strength is much less.

Quarantine treatments strong enough to cause very high levels of mortality (such as probit 9) may also cause some commodity damage or reduce product shelf life. Lower efficacy treatments, if justified, could reduce commodity damage and maintain shelf life. In turn, this could reduce costs for exporters and result in a more attractive, longer lasting product for consumers. Systems must be in place to ensure that the allowable commodity infestation level is not exceeded.

Reduce p, μ and/or ϕ with a Specific Host Cultivar

Suppose a new cultivar of the commodity had been developed that was less preferred by the pest as a host and less suitable for development when infested. Trial data show that $p = 0.004$, $\mu = 2.4$, and $\phi = 0.1$. The value of Φ, using Equation 4.2, is now 6.4×10^{-3}, resulting in a value of r, using Equation 4.1, of nearly one. The new cultivar has reduced the probability of establishment, but not enough, and it is decided to use a post-harvest treatment as well.

To determine the efficacy of the required treatment, we must first determine the required values of Φ and ϕ. With an infestation level of 0.004 and r = 0.0001, Φ must be 5×10^{-8} (Equation 4.5). Solving Equation 4.2 for ϕ gives a value of 0.00026. The required post-harvest treatment efficacy is $0.00026 \div 0.1$, or 0.0026 equivalent to probit 7.8. In this example, the use of the resistant cultivar has an equivalent effect to the use of the pre-export clearance data above. Appropriate quarantine actions could include the use of the resistant cultivar and treatment of probit 7.8, or an infestation level of 3×10^{-5} (based on pre-export clearance data) and treatment of probit 7.8. Thus, this model can be used to demonstrate the equivalence of different quarantine actions.

Reduce Ψ

Assume that the new cultivar was to be imported into a cooler portion of the country, during winter. Re-assessment of the suitability of prevailing climatic conditions for pest establishment now results in a value of 0.1 for Ψ.

The value of Φ for the new cultivar, using Equation 4.2, is now 0.0013, making $r = 0.92$ (Equation 4.1). Again, the reduction in Ψ has reduced r, but not by enough. To determine the efficacy of a suitable post-harvest treatment, set r in Equation 4.5 to 0.0001 and solve for Φ. The necessary value is 5×10^{-8}. Solve for the value of ϕ that results in $\Phi = 5 \times 10^{-8}$ in Equation 4.2 yields $\phi = 0.00059$, resulting in a post-harvest treatment efficacy of $0.00059 \div 0.1$ or 0.0059, equivalent to probit 7.5.

Reduce N

Using Equation 4.1, reducing N is a mathematical, if not commercially viable, method of reducing r. Suppose it was decided to set the probability of establishment from each possible entry pathway, commercial and non-commercial, at 0.0001. Since restricting trade (by reducing N) is not an option, methods such as those discussed above would have to be used for the commercial consignment of 500,000 units. However, reducing N is an option for meeting the target probability from non-commercial pathways such as passenger baggage.

For non-commercial pathways, no assumptions about specific cultivars can be made, although passenger arrival data can be used to reassess Ψ if the season or location of passenger arrivals differ significantly from cargo. If border interception data exist, it may be possible to determine an infestation level for the commodity arriving in passenger baggage, otherwise, a value of 1 should be used.

Using the initial data for the commodity, the probability of pest establishment, Φ, is 0.3 as determined from Equation 4.2. This is the probability that a single infested item causes an establishment. Since no information is known about p, any item slipping through could be infested, meaning that even if $N = 1$, the probability of establishment, r, would be 0.3 (Equation 4.1). Thus, not even one unit slipping through in passenger baggage would be acceptable, given the standard of $r = 0.0001$.

Reconciling commercial pathway risk with risk from non-commercial pathways is an area that requires further development. However, this model can be used to determine equivalent standards for multiple pathways.

Summary and Conclusions

This chapter has introduced a methodology for quantifying technical trade barriers that contain elements of risk of occurrence and economic impacts, and for creating benchmarks and comparing quarantine treatments. The important change from previous practice is that both economic and scientific criteria can be included in the analysis, ensuring that barriers can be treated on the basis of expected outcome rather than the technical consideration of biological factors only. As such, it is possible to step beyond considering whether the barrier involves an insect or bacteria, and instead focus on whether a potential event behind the barrier is above, below, or within an expected dollar value.

The iso-risk framework deals with some of the problems created by SPS in a trade environment, such as the even treatment of technical barriers and the need for transparent and measurable criteria for justifying decisions to trading partners. Using iso-risk, equivalent treatment requires that technical barriers or SPS have similar outcomes. This means that two exporters can be subjected to different quarantine requirements but not violate GATT rules on equal treatment, since the outcomes of the measures are similar. Justification of quarantine measures also becomes easier, since decisions can be shown to be consistent within an overall domestic policy context.

The key components of the iso-risk framework, economic impact and probability of establishment, are used in general terms in this chapter, since the purpose here is to outline a methodology rather than define specific standards. The U.N.'s FAO guidelines on Pest Risk Analysis allow for economic impacts ranging from partial budget analyses of impacts on producers to general equilibrium models which look at economy-wide impacts from changing product prices and factor costs. However, these guidelines leave it to individual countries to determine which model to use (FAO, 1996). The FAO guidelines also outline the factors that should be considered in determining risk of establishment, but do not specify in detail how this should be done. It is important to recognize that the choice of models for determining economic impact or probability of establishment is not critical to the methodology described in this chapter. What is critical is how these two factors are combined and interpreted.

To develop standards by which quarantine measures might be objectively compared, some consensus will be required on the appropriate economic impact to measure and the calculation of risk of establishment. In the early stages of establishing iso-risk, a country would only be able to determine whether it is treating its trading partners consistently using domestic definitions of economic impact and probability of establishment. This internal consistency of quarantine policy would be relative to a domestic ALP. At a later stage, when a number of countries would base decisions on iso-risk, it is possible that an international norm for ALP would emerge. A country could then establish treatment of trading partners consistent with international norms, or be challenged to do so.

The problem of arriving at an ALP which adequately describes a regulatory agency's perception of acceptable pest risk in an iso-risk framework can be approached by starting with a country's current regulatory treatment of pests and commodities. To establish an ALP, a sufficient sample of pests would first need to be evaluated for probability of entry and potential economic impacts. ALP should emerge from the pattern of plotted results, represented by a line above which there would be no plots. A similar process could determine an ALP for commodities. The only difference would be that commodities, rather than individual pests, would be evaluated. A value for ALP implicit in existing quarantine regulations should emerge from the analysis. The process is not likely to be easy in practice, since such an analysis may show inconsistencies in existing quarantine policies based on the resulting values of commodity and pest risk.

While providing a clearer picture of ALP, experience in New Zealand has shown that there is a significant increase in information and analysis required by a quarantine authority when it has to include an economic impact assessment and a specific probability of introduction. In many cases, little will be known about the economics of particular crops, much less the expected economic impact on a particular plant, or probabilities of introduction. In addition to the problem of basic data, there is a problem with producing a rapid analysis for quarantine decisions if the level of detail implied by iso-risk is required for each commodity traded. Models to facilitate rapid analysis have been developed for MAF that calculate probability of introduction (Whyte et al., 1996; Baker et al., 1993) and economic impacts (Bigsby and Crequer, 1995; Bigsby, 1995b) based on a standardized set of factors, and work has been progressing to develop a database for risk assessment.

References

Baker, A.C., 1939, The basis for treatment of products where fruitflies are involved as a condition for entry into the United States, USDA Circular 551, Washington, D.C.

Baker, R.T., Cowley, J.M., and Harte, D.S., 1993, Pest risk assessment. A process for the assessment of pest risk associated with the importation of plants and plant products into New Zealand, *Lynfield Plant Protection Centre Publications, No. 1,* New Zealand Ministry of Agriculture and Fisheries, 16 pp.

Bigsby, H.R., 1996a, *Enhancement and Further Development of the Economic Impact Assessment Modules,* final report for MAF Regulatory Authority, Wellington.

Bigsby, H.R., 1996b, Quantifying Technical Trade Barriers: Phytosanitary Measures, proceedings of the NZAES Conference, Blenheim, July 5–6, 1996, 124–128.

Bigsby, H.R., 1995a, Estimating the Economic Impacts of Pest Introduction, Market Effects and Partial Equilibrium Analysis, AERU report for the Ministry of Agriculture and Fisheries, Wellington.

Bigsby, H.R., 1995b, Pest Model—Partial Equilibrium Model User's Manual, AERU report for the Ministry of Agriculture and Fisheries, Wellington.

Bigsby, H.R., 1994a, Estimating the Economic Impact of Horticultural Imports and Associated Pest Risk, paper presented at the International Plant Protection Convention (IPPC) Working Group Meeting, Ottawa, Canada, Nov. 15–17, 1994.

Bigsby, H.R., 1994b., Economic Impacts of Horticultural Imports and Associated Pest Risk, report prepared for Ministry of Agriculture and Fisheries Regulatory Authority, Wellington.

Bigsby, H.R. and Crequer, J., June, 1996, A Conceptual Model for the Management of Pest Risk, paper presented at the SEEM 2 Conference, Decision Making and Risk Assessment in Biology, University of Otago.

Bigsby, H.R. and Crequer, J., 1995, Pest Plant Economic Impact Assessment Model, prepared for MAF Regulatory Authority, Wellington.

Bigsby, H.R. and Whyte, C., 1998, A Model of the Appropriate Level of Protection for New Zealand's Quarantine Security, MAF Policy Technical Paper, Wellington.

Cohrssen, J.J. and Covello, V.T., 1989, Risk analysis: a guide to principles and methods for analyzing health and environmental risks, United States Council on Environmental Quality, Executive Office of the President, 407.

Couey, H.M. and Chew, V., 1986, Confidence limits and sample size in quarantine research, *J. Econ. Entomol.,* 79, 887–890.

FAO, 1996, International Standards for Phytosanitary Measures, International Plant Protection Convention, FAO, Rome.

Greer G. and Bigsby, H.R., 1995, Estimating the Economic Impacts of Pest Introduction, AERU report for the Ministry of Agriculture and Fisheries, Wellington.

Greer G., Bigsby, H.R., and McAuliffe, R., 1995, Pest Model—Partial Budget Model User's Manual, AERU report for the Ministry of Agriculture and Fisheries, Wellington.

Hillman, J., 1991, Technical Barriers to Agricultural Trade, Westview Press,

Boulder.

Hillman, J., 1978, Nontariff Agricultural Trade Barriers, University of Nebraska Press, Lincoln.

Kaplan, S. and Garrick, B.J., 1981, On the quantitative definition of risk, *Risk Analysis 1,* 11–27.

Miller, L., McElvaine, M.D., McDowell, R.M., and Ahl, A.S., 1993, Developing a quantitative risk assessment process, *Rev. Sci. Tech. Off. Int. Epiz., 12,* 1153–1164.

Ministry of Health, 1996, Managing risks—a concept paper, Report of the Food & Nutrition Section, Public Health Group, New Zealand Ministry of Health, 30.

Ndayisenga, F. and Kinsey, J., 1994, The structure of nontariff trade measures on agricultural products in high-income countries, *Agribusiness* 10(4), 275–292.

Orr, R.L., 1995, A plant quarantine risk-based framework for standardizing the acceptable level of risk and appropriate level of protection, Animal and Plant Health Inspection Service, USDA.

Papasolomontos, A., 1993, Present global activities in the harmonization of plant quarantine, in *International Approaches to Plant Pest Risk Analysis,* proceedings of the APHIS/NAPPO International Workshop on the Identification, Assessment and Management of Risks due to Exotic Agricultural Pests, Alexandria, VA, NAPPO Bulletin 11.

Patterson, E., 1990, International efforts to minimize the adverse trade effects of national sanitary and phytosanitary regulations, *J. World Trade,* 24(2), 91–102.

Smith, J.R., 1993, Welcoming remarks, in *International Approaches to Plant Pest Risk Analysis,* Proceedings of the APHIS/NAPPO International Workshop on the Identification, Assessment and Management of Risks due to Exotic Agricultural Pests, Alexandria, VA, NAPPO Bulletin 11.

Whyte, C.F., Baker, R.T., Cowley, J.M., and Harte, D.S., 1996, Pest establishment. A quantitative method for calculating the probability of pest establishment from imported plants and plant products, as a part of pest risk assessment, *NZ Plant Protection Centre Publications No. 4,* New Zealand Ministry of Agriculture, 11.

U.S. Department of Agriculture (USDA), 1995, Pathway-Initiated Pest Risk Assessment: Guidelines for Qualitative Assessments, version 4.0, USDA-APHIS-PPQ, Riverdale, MD.

Chapter 5

Food Safety Issues in Developing Nations: A Case Study of Brazil

Elisabete Salay, José Luiz Pereira, Adriana Zenlotti Mercadante, Flávia Maria Netto, and Suzi Barletto Cavalli

Many problems related to food safety are being noticed in developing countries; however, multidisciplinary studies on this topic are extremely rare. In this chapter, three case studies in Brazil will be evaluated: the occurrence of mycotoxins in foods, the occurrence of foodborne illness outbreaks, and technological constraints in the cold-chain infrastructure. A multidisciplinary approach was used involving policy development, agricultural economics, food chemistry, food microbiology, and food quality. Although there is no efficient information system yet on the number of diseases linked to food in Brazil, the results of this research suggest that the Brazilian population is facing important health risks arising from food, such as food contamination with mycotoxins. A higher frequency of foodborne diseases was related to poor hygienic and sanitary conditions. It was verified that the cold chain in Brazil has qualitative problems, particularly in its geographic dispersion. The government has implemented various regulations to control food safety and continues to follow such reforms. Even though the number of enterprises concerned with food safety in Brazil is growing, this sector still faces many problems specific to developing countries.

0-8493-2217-0/01/$0.00+$.50

Introduction

The system of food safety control in developing countries is extremely complex. Different levels of technological development can be found in these food systems, such as modern and traditional processes which range from rigorous quality control to no control at all. These food control systems aim to solve safety problems ranging from biotechnologically-transformed products to common hygienic practices. However, studies which aim for an understanding of the true magnitude of food safety problems in developing countries are very limited.

This chapter contributes to this theme by analyzing three case studies in Brazil. Despite having a Gross National Product (GNP) of 778 billion U.S. dollars in 1998, and now being classified by the World Bank as an upper middle class income country (World Bank, 1999), Brazil suffers one of the worst national income distributions in the world, giving rise to serious social problems. The net rate of secondary education enrollment in the country is still only 20%. Access to potable water is restricted to 69% of the population, with reliable sanitation available to only 67% (World Bank, 1999). Significant differences in economic and social development are observed throughout the country.

The agricultural and animal husbandry sectors play a fundamental role in the Brazilian economy, since together they support a trade surplus of 6.5 billion U.S. dollars in 1998 (Associação Brasileira da Indústria de Alimentação, 1999a). Furthermore, the physical production of the food industries increased 18.5% from 1994 to 1998, representing 9.8% of the GNP by 1998 (Associação Brasileira da Indústria de Alimentação, 1999b). Despite this impressive improvement in food production capacity, certain poorer segments of the population still fail to satisfy their food needs (Lavinas et al., 1998).

The guarantee of food safety for a population is considered by many to be the shared responsibility of government policies, the activities of educated and organized consumers, and responses to incentives by agribusinesses. In Brazil, Ministries of Health and Agriculture and Supply (MAS), which regulate and inspect food establishments and their products, are responsible for public policies on food safety (Salay and Caswell, 1998). Few evaluations of these public interventions have been made from the socioeconomic point of view. Also, little is known of the importance placed on food safety activities within the strategies of national firms.

This chapter presents information on specific topics concerning Brazil, using a multidisciplinary approach. An interdisciplinary group was formed to assess the current status of food safety in Brazil. This group consisted of specialists in the areas of policy development, agricultural economics, food science, food quality, and food microbiology. An interdisciplinary analysis of food safety in Brazil is practically nonexistent. It was decided that the most convenient initial approach would be to diagnose the extent of food safety problems in the country. From this starting point, case studies aimed at identifying the preferable manner of government action for specific foods were selected, and the problems of incentives for firms to adopt enhanced methods to improve the level of food safety were developed. The role of consumers in food control is still under investigation and will not be included in this chapter. The respective case studies described are the occurrence of mycotoxins in foods and the case of aflatoxin in peanuts, the occurrence of foodborne illness outbreaks and the case of safety in food service operations, and the cold chain[1] in Brazil and the issue of milk refrigeration. These case studies utilize data on chemical agents such as mycotoxins, as well as microbiological food safety hazards. The geographic dispersion and related technical problems of the cold chain are also discussed.

Assessing the Problems with Case Studies

Mycotoxins

Introduction

Mycotoxins are secondary fungal metabolites that can produce harmful effects in humans and animals. More than 300 of these toxic compounds are known, although the effects of only a few have been well evaluated. They present a wide variety with respect to chemical structure and toxic effects (Sabino and Rodriguez-Amaya, 1993).

Mycotoxin contamination may occur in the field when toxigenic fungi are present and the grains are exposed to drought, insect damage, unusually high rain levels or, in some cases, excessively low temperatures. Improper storage conditions have been associated as one of the main causes of mycotoxin contamination. The fungi that most frequently colonize grains in the field are *Alternaria* and *Cladosporium,* which seldom develop further during storage. The species predominantly found during storage are *Aspergillus* and *Penicillium*. *Fusarium* species can develop in both stages (Christensen and Kaufmann, 1969).

Aflatoxins, produced exclusively by *Aspergillus flavus* and *A. parasiticus,* cause the greatest damage to humans and animals, since they occur widely and are highly toxic. The most important aflatoxins found in foods, B_1 (Figure 5.1), G_1, B_2, and G_2, show carcinogenic, mutagenic, and

[1]The cold chain includes a series of refrigeration operations which vary from product to product and are characteristic of each type of marketing strategy (Pinazza and Alimandro, 1999).

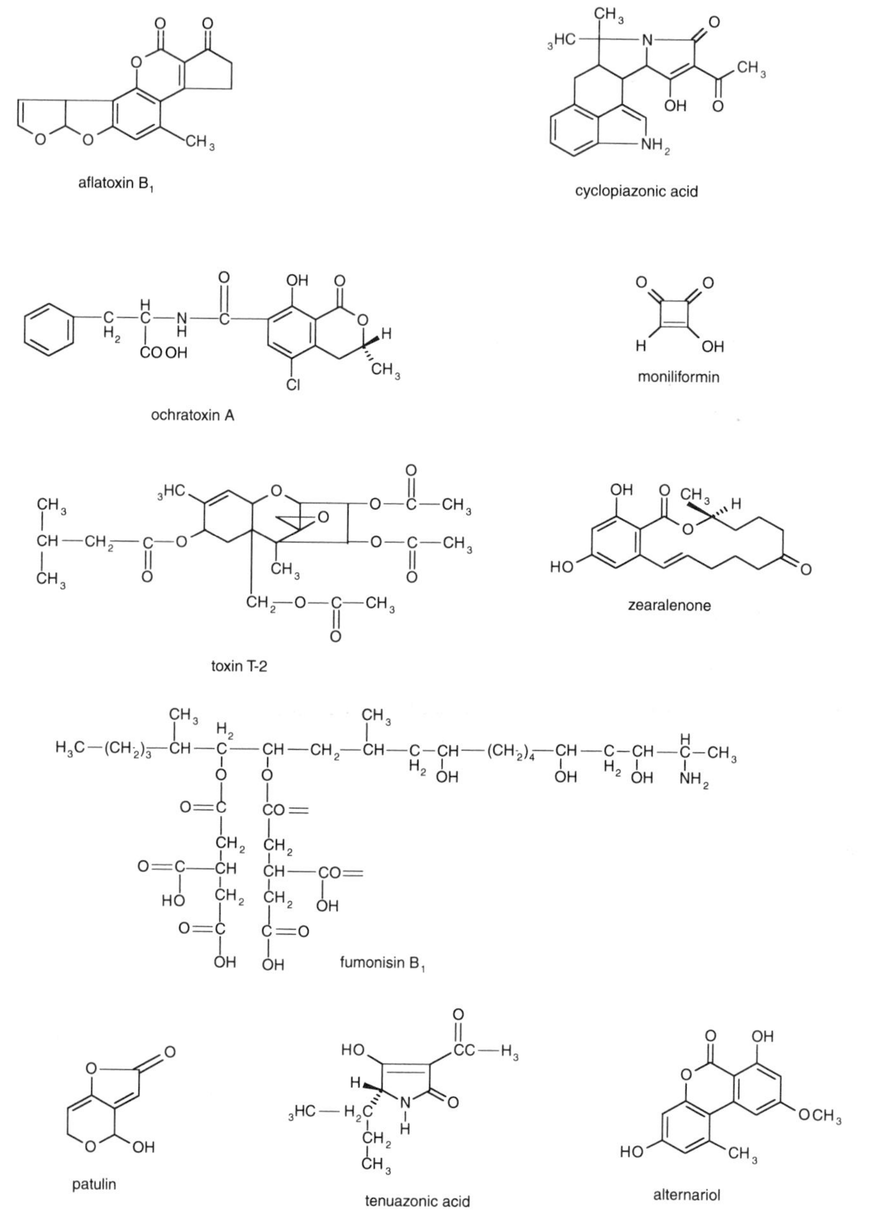

Figure 5.1 Structures of Mycotoxins Commonly Found in Foods

teratogenic properties in humans and many animals (Ellis et al., 1991). Aflatoxin M_1 is a hepatocarcinogen found in the milk of animals that have consumed feed contaminated with aflatoxin B_1 (Van Egmond, 1989).

Cyclopiazonic acid and ochratoxins (Figure 5.1) are produced by several species of *Aspergillus* and *Penicillium,* and both can coexist with aflatoxins. A variety of symptoms such as weight loss, diarrhea, depression, convulsions, and death, as well as degeneration and necrosis of muscles and viscera, were observed in animals contaminated with cyclopiazonic acid (Pollock et al., 1982). Ochratoxin A, the most common of the ochratoxins, is nephrotoxigenic and teratogenic, the main dangerous effects being necrosis of the renal tubular epithelium and the periportal hepatic cells (Pohland et al., 1992).

Trichothecenes are toxic metabolites produced mainly by the *Fusarium* species. Those most commonly found in grains are toxin T-2 (Figure 5.1) and its derivatives, and deoxynivalenol. The toxic acute effects produced in humans are characterized by vomiting, diarrhea, anorexia, hematologic changes, neurological disturbances, destruction of the bone marrow, and generalized hemorrhaging, which may or may not be followed by death. The diversity of symptoms observed indicates that the trichotecenes act at the transcription and translation stages of RNA. Decomposition of trichothecenes may occur during grinding, malting, and fermentation, as well as during the production of bread, macaroni, and beer (Scott, 1984; Snyder, 1986). *Fusarium* may also produce the teratogenic compound zearalenone (Figure 5.1) (Kuiper-Goodman et al., 1989).

Some species of *Fusarium,* especially *F. moniliforme,* produce moniliformin and fumonisins. Equine leukoencephalomalacia is the main disease associated with fumonisins B_1 (Figure 5.1) and B_2, although evidence of esophageal cancer in humans has also been found (Scott, 1993).

Different species of *Penicillium, Aspergillus,* and *Byssochlamis* are able to produce patulin (Figure 5.1), which shows antibiotic properties against many bacteria and fungi. However, patulin also shows high levels of toxicity to plants and to the cells and tissues of animals (Engel and Teuber, 1984). This mycotoxin can be produced during the refrigerated and nonrefrigerated storage of apples.

The genus *Alternaria,* which frequently develops in tomatoes, can produce the mycotoxins tenuazonic acid, alternariol, and alternariol monomethyl ether, which show weakly acute effects.

Current Brazilian legislation has fixed a maximum permitted level of 20 ppb (parts per billion) for the sum of the aflatoxins $B_1 + B_2 + G_1 + G_2$ in foods for human consumption. However, there is no legislation for other mycotoxins in Brazil.

Occurrence

The various surveys on the occurrence of mycotoxins conducted in Brazil during the period 1990 to 1999 are summarized in Table 5.1. These data were collected from published research articles found in the database of Food Science and Technology Abstracts. Abstracts presented at the National Scientific Congress on Mycotoxins (Encontro Nacional de

Table 5.1 Occurrence of Mycotoxins in Brazil during the Period 1990 to 1999

					Contaminated Samples			
Product	*State*	*Number of Samples*	*Period of Collection*	*Mycotoxins Detected*	*Mycotoxin*	*Level (ppb)*	*Number of Samples*	*Reference*
Peanuts	SP	1,115	1990–1996	AFLA B_1, B_2, G_1, G_2	AFLA B_1, B_2, G_1, G_2	63–948	658	Fonseca et al. (1998)
Peanuts and corn flour	RS	120	1991	AFLA B_1, B_2, G_1, G_2	AFLA B_1, B_2, G_1, G_2	10–805	86	Baldissera et al. (1992)
Peanuts and products, corn flour and popcorn	SP	53	1992	AFLA B_1, B_2, G_1, G_2, CPA,	AFLA B_1, B_2, G_1, G_2, CPA,	8–320	11	Sylos and Rodriguez-Amaya (1994)
Peanuts	PE	86	1993	AFLA B_1, B_2, G_1, G_2	AFLA B_1, B_2, G_1, G_2	10–2,000	26	Araújo et al. (1994)
Peanuts and products	SP	321	1994	AFLA B_1, B_2, G_1, G_2	AFLA B_1, G_1	> 30	116	Sabino et al. (1996)
Peanuts and products	SP	66	1994	AFLA B_1, B_2, G_1, G_2	AFLA B_1, B_2, G_1, G_2	14–997	32	Brigido et al. (1995)
Peanuts	PE	108	1996–1997	AFLA B_1, B_2, G_1, G_2	AFLA B_1, G_1	14–1,178	46	Oliveira et al. (1998)
Peanut, peanut products, corn	SP, PR, SC, RS	131	—	AFLA B_1, B_2, G_1, G_2, CPA	AFLA B_1, G_1, CPA	8–2,152	54	Sylos et al. (1996)

Peanuts, wheat flour, beans, corn	SC	56	—	AFLA B_1, B_2, G_1, G_2, OCHRA, ZEA, STERI	AFLA B_1, B_2, G_1, G_2	30–356	4	Costa and Scussel (1998b)
Peanuts and products, rice, beans, corn and cassava flour, popcorn	SP	113	—	AFLA B_1, B_2, G_1, G_2, OCHRA, ZEA, STERI	AFLA B_1, B_2 G_1, G_2	6–118	16	Sylos (1998)
Popcorn	SP	227	1988–1990	AFLA B_1, B_2, G_1, G_2, OCHRA, ZEA, STERI	AFLA B_1, B_2, G_1, G_2, ZEA	1–462	16	Soares and Furlani (1992)
Corn	SP	130	1991	AFLA B_1, B_2, G_1, G_2, OCHRA, ZEA, STERIG, DON, fumonisin B_1	AFLA B_1	500	1	Pozzi et al. (1995)
Freshly harvested corn	SP	16	1992	AFLA B_1, B_2, G_1, G_2, OCHRA, ZEA	—	—	0	Castro et al. (1995)

Table 5.1 *(continued)*

Product	State	Number of Samples	Period of Collection	Mycotoxins Detected	Contaminated Samples: Mycotoxin	Level (ppb)	Number of Samples	Reference
Corn	PR	816	1992–1994	AFLA B_1, B_2, DON, ZEA, fumonisin B_1, T-2	AFLA B_1, B_2, DON, ZEA, fumonisin B1 T-2	0.7–4,600	210	Lazzari (1994)
Corn	PR	150	1995–1996	AFLA B_1	AFLA B_1	11.6–7,400	17	Ono et al. (1998a)
Products of corn, rice and wheat	RS	165	1996	AFLA B_1, B_2, G_1, G_2, OCHRA, ZEA	AFLA B_1, OCHRA, ZEA	—	10	Badiale-Furlong et al. (1998)
Corn products	RS	213	1997	AFLA B_1, B_2, G_1, G_2	AFLA B_1	3.2–25.6	37	Pich et al. (1998)
Popcorn	CE	70	1991	AFLA G_1	AFLA B_1	22–130	23	Vale (1992)
Corn	SP	35	1994/95	fumonisins B_1, B_2	fumonisins B_1, B_2	30–6,160	35	Camargos et al. (1998)
Corn	SP	42	1994–1995	fumonisins B_1, B_2	fumonisins B_1, B_2	130–12,250	40	Machinski et al. (1998)
Corn	PR, MS, GO	48	1990/91	fumonisins B_1, B_2	fumonisins B_1, B_2	600–19,130	47	Hirooka et al. (1996)

Freshly harvested corn	PR	113	—	fumonisins B_1, B_2	fumonisins B_1, B_2	10–10690	111	Ono et al. (1998b)
Corn	SP	22	1992	moniliformin	—	—	0	Leoni et al. (1994)
Freshly harvested corn	PR, GO	88	1994/95	DON, T-2	DON, T-2	104–125	6	Prado et al. (1997)
Wheat	RS	12	1988–1990	AFLA B_1, B_2, G_1, G_2, OCHRA, ZEA, STERI, DON, NIV, DAS, T-2, HT-2, T-2 triol, T-2 tetraol	OCHRA, DON	0–0.4	2	Furlong et al. (1995)
Wheat flour	RS	33	1995	AFLA B_1, B_2, G_1, G_2, OCHRA, ZEA, STERI	—	—	0	Vieira and Badiale-Furlong (1998)
Wheat flour, bread, pizza	RS	272	1996–1997	AFLA B_1, B_2, G_1, G_2, OCHRA, ZEA,	OCHRA, ZEA	26–53	0	Vieira et al. (1998)

Table 5.1 *(continued)*

Product	State	Number of Samples	Period of Collection	Mycotoxins Detected	Contaminated Samples			Reference
					Mycotoxin	Level (ppb)	Number of Samples	
Nuts (almonds, cashew nuts, Macadamia nuts, Brazil nuts, walnuts, hazelnuts)	SP	110	1991 and 1995	AFLA B_1, B_2, G_1, G_2, OCHRA, ZEA, STERI	AFLA B_1, G_1	10–26	2	Furlani and Soares (1996)
"Natural products", cereal breakfast	SP	69	1991	AFLA B_1, B_2, G_1, G_2, OCHRA, ZEA, STERI	—	—	0	Soares and Furlani (1996a)
Wheat sold in health stores	SP	38	1991	AFLA B_1, B_2, G_1, G_2, OCHRA, ZEA, STERI, DON, NIV, DAS, T-2, HT-2, T-2 triol, T-2 tetraol	—	—	0	Soares and Furlani (1996b)

Beans	SC	72	1997	AFLA B_1, B_2, G_1, G_2, OCHRA, ZEA, STERI	AFLA B_1, B_2, G_1, G_2, OCHRA	0–53	3	Costa and Scussel (1998a)
Green coffee beans	PR, SP, MG, ES, RO, BA	50	—	OCHRA	OCHRA	0.8–117.4	15	Furlani et al. (1998)
Pasteurized milk, powdered milk, cheese, yogurt	SP	204	1989/90 and 1992	AFLA M_1	AFLA M_1	73–370	4	Sylos et al. (1996)
Milk	SP	144	1992–1993	AFLA M_1, M_2	—	—	0	Corrêa et al. (1996)
Cheese	SP	36	—	AFLA B_1, B_2, G_1, G_2, M_1, OCHRA, patulin, citrinin, penicillic acid	—	—	0	Taniwaki and Van Dender (1992)
Powdered milk	SP	60	—	AFLA M_1	—	—	0	Navas et al. (1994)
Powdered milk	SP	300	1992–1993	AFLA M_1	AFLA M_1	0.1–1.0	33	Oliveira et al. (1997)

Table 5.1 *(continued)*

Product	State	Number of Samples	Period of Collection	Mycotoxins Detected	Contaminated Samples: Mycotoxin	Contaminated Samples: Level (ppb)	Contaminated Samples: Number of Samples	Reference
Fruit and fruit juice	SP, SC, PR, RS	149	1992, 1993, 1995	patulin	patulin	17	1	Sylos and Rodriguez-Amaya (1999)
Apple juice	PR	73	1992–1993	patulin	patulin	6.4–77.5	15	Machinsky and Midio (1996)
Tomato products	SP	80	—	AOH, AME, TEA, CPA	TEA, CPA	34–177.9	19	Motta and Soares (1996)
Liver paste	SP	40	—	AFLA B_1, B_2, G_1, G_2, M_1	AFLA B_1	2.3	1	Rosa et al. (1996)
Eggs	RJ	45	—	AFLA B_1, M_1,l aflatoxico	AFLA B_1	2.2–4.9	2	Fraga et al (1994)

BA: Bahia, CE: Ceará, ES: Espírito Santo, GO: Goiás, MG: Minas Gerais, MS: Mato Grosso do Sul, PE: Pernambuco, PR: Paraná, RJ: Rio de Janeiro, RO: Rondônia, RS: Rio Grande do Sul, SC: Santa Catarina, SP: São Paulo, AFLA: aflatoxin; CPA: cyclopiazonic acid; OCHRA: ochratoxin, ZAE: zearalenone; STERI: sterigmatocystin; DON: deoxynivalenol; NIV: nivalenol; DAS: diacetoxyscirpenol; AOH: alternariol; AME: alternariol monomethylether; TEA: tenuazonic acid.

Micotoxinas) and at a Latin American Symposium on Food Science, each of which are held every two years in Brazil, and at the IX International Symposium on Mycotoxins and Phycotoxins (sponsored by International Union of Pure and Applied Chemistry, 1996), were also considered.

The data presented in Table 5.1 show that about 52% (969 samples) of peanuts and peanut products were contaminated with aflatoxins. Of a total of 1866 samples, 28% had levels of aflatoxins B_1 plus G_1 greater than 30 ppb, the maximum level tolerated by the Brazilian legislation up to March 1996. Due to climatic variations, the frequency and level of contamination changes drastically depending on the year of harvest. In fact, several Brazilian studies conducted in the 1960s showed that the incidence of aflatoxin was higher in peanuts harvested during the rainy season (Fonseca, 1968; Tango et al., 1966), probably as a result of poor drying.

In the only two studies available (Sylos and Rodriguez-Amaya, 1994; Sylos et al., 1996) that investigated cyclopiazonic acid, 11 samples of peanuts and their products, from a total of 79 samples, were contaminated with this mycotoxin in the range of 150 to 369 ppb. Aflatoxin was also found in most of these contaminated samples. These results confirm that the contamination of peanuts in São Paulo State, which is the major producer of these commodities, is still a serious problem.

Corn is constantly used, mixed with other ingredients for animal feeding, and also a staple food for humans in some Brazilian regions. The consequences in humans of constantly consuming corn naturally contaminated with mycotoxins have not been completely elucidated. Of a total of 227 samples of corn and its products produced in São Paulo State (Castro et al., 1995; Pozzi et al., 1995; Soares and Furlani, 1992; Sylos and Rodriguez-Amaya, 1994; Sylos, 1998), only 2 samples (0.9%) were contaminated with aflatoxin B_1. However, 26% of 511 samples of corn, both grain and processed, cultivated or produced in the southern states (Paraná, Santa Catarina, and Rio Grande do Sul) were contaminated with aflatoxins (Badiale-Furlong et al., 1998; Lazzari, 1994; Ono et al., 1998a; Pich et al., 1998; Sylos et al., 1996). Cyclopiazonic acid was only detected in samples (14%) originating from the south (Sylos et al., 1996). The contamination with aflatoxins was also low (6%) for popcorn produced in São Paulo (Soares and Furlani, 1992; Sylos and Rodriguez-Amaya, 1994; Sylos, 1998), while the incidence was higher (33%) when the popcorn was from a warmer state, Ceará (Vale, 1992).

On the other hand, the incidence of contamination of corn and its products with fumonisins seems to be widespread. Almost all (98%) of the corn samples cultivated in São Paulo State and analyzed during the years 1994 and 1995 were contaminated with fumonisins B_1 and B_2 (Camargos et al., 1998; Machinski et al., 1998). The same tendency was observed in corn from Paraná, Mato Grosso do Sul, and Goiás (Hirooka et al., 1996; Lazzari, 1994; Ono et al., 1998b).

Despite the high incidence of fumonisins in Brazilian corn, moniliformin, which is also produced by *Fusarium,* was not detected in 22 samples from 20 corn-producing cities in São Paulo State (Leoni et al., 1994). The incidence of toxin T-2 and deoxynivalenol was also low (7%) in 88 corn samples produced in the states of Paraná and Goiás (Prado et al., 1997). However, despite the low incidence of these mycotoxins, it must be remembered that the number of samples analyzed was very small, and therefore these studies should continue.

The absence of aflatoxin was observed in 12 samples of wheat (Furlong et al., 1995), 166 samples of wheat flour (Badiale-Furlong et al., 1998; Vieira and Badiale-Furlong, 1995; Vieira et al., 1998), and in 218 wheat products (Vieira et al., 1998). Trichothecenes were also not detected in 12 samples of wheat (Furlong et al., 1995), and the contamination of wheat flour with zearalenone and ochratoxin was very low (less than 5% of the total number of samples). A low incidence (2%) of the samples, with levels higher than those accepted by Mercosur countries (Brazil, Argentina, Uruguay, and Paraguay) of aflatoxins B_1, B_2, G_1, and G_2, ochratoxin, zearalenone, and sterigmatocystin in nuts (Furlani and Soares, 1996) has been reported, as well as negative results for contamination with the above mycotoxins in natural products sold in health stores (Soares and Furlani, 1996a, b) and breakfast cereals (Soares and Furlani, 1996b). Of a total of 170 samples of rice and beans, commodities consumed daily in Brazil, as well as samples of rice and cassava flour, 3% were contaminated with aflatoxins and ochratoxin A at levels lower than 53 ppb (Badiale-Furlong et al., 1998; Costa and Scussel, 1998a, b; Sylos, 1998).

These data indicate that the occurrence of aflatoxins in all kinds of food and agricultural products is not as widespread in the tropics as generally believed. Moreover, sterigmatocystin was not found in any of Brazilian commodities analyzed (Costa and Scussel, 1998a, b; Furlani and Soares, 1996; Furlong et al., 1995; Pozzi et al, 1995; Soares and Furlani, 1992; Soares and Furlani, 1996a, b; Sylos, 1998; Vieira and Badiale-Furlong, 1998).

Only one study of coffee has been conducted in Brazil (Furlani et al., 1998). About 30% of the green coffee for the Brazilian market was contaminated with ochratoxin A, indicating that a greater number of samples should be analyzed in order to give a better picture of this problem.

The presence of aflatoxin M_1 in milk and cheese is a worldwide concern, especially as predominantly children consume milk. The analysis of a total of 380 samples of different types of pasteurized milk and cheese marketed in São Paulo State (Corrêa et al., 1996; Sylos et al., 1996; Taniwaki and Van Dender, 1992) indicated that the incidence of aflatoxin M_1 (4 samples contaminated) is not serious. A possible explanation is that cows in this region graze year round. Although only 2% of milk powder samples (Navas et al., 1994; Oliveira et al., 1997) showed contamination above the tolerance limit established by the Food and Drug Administration (Stoloff et al., 1991), it appears that contamination is more common in powdered milk because of the removal of water during the process.

Since patulin is associated with spoiled fruits and is only partially destroyed by processing, the patulin content of juice may be used as a good indicator of the quality of the fruits used as inputs (Sylos and Rodriguez-Amaya, 1999). From the results presented in Table 5.1, patulin does not seem to be a problem in fruit juices marketed in the states of São Paulo and Paraná (Machinsky and Midio, 1996; Sylos and Rodriguez-Amaya, 1999), since only one sample was contaminated above the maximum limit, 50 μg/l, established by the World Health Organization. The low incidence of patulin in Brazilian juices can be explained by the addition of the preservative sulfur dioxide, permitted by Brazilian regulation, which was reported to decrease patulin concentration (Burroughs, 1977).

Peanuts

On June 10, 1997, by way of Decree No. 230, the Ministry of Agriculture and Supply created the National Program for the Control of Mycotoxins in products, sub-products, and derivatives of vegetable origin, with power to develop actions for the education, monitoring, and inspection of pre- and post-harvest technologies of these products (Ministério da Agricultura e do Abastecimento, 1997). The program was designed to develop legislation, laboratories, training and allocation of human resources, monitoring, research, and rural extension (Vargas, 1998). The Ministry of Health is also developing standards for mycotoxins in foods in Brazil. Other interventions developed prior to this decree concentrated on program structure such as the certification of laboratories, standardization of analytical methodologies, and training. Before November 1999, inspection by the public sector using analyses for mycotoxins had not begun, despite advanced planning due to lack of financial resources. MAS is currently aiming at utilizing programs of technical cooperation to assist with the expenses.[2]

According to MAS (Souza et al., 1998), 1345 and 1185 products were analyzed for aflatoxin in 1996 and 1997, respectively, by 5 laboratories certified by MAS. All these analyses were paid for and requested by producers. The results confirmed the previously observed tendency for peanuts and peanut products to be the most contaminated products with respect to aflatoxin. Thus, it can be seen that the public control of mycotoxins in foods is still not 100% effective. A revised edition of this decree redefining the program is to be published by the end of 2000.

---

[2]Information obtained from Ministry of Agriculture and Supply.

The constant problem of contamination of Brazilian peanuts with mycotoxins has blocked their exportation to most countries. Thus, the majority of the production of this oleaginous seed is destined for the domestic market for human consumption, either unprocessed or in confectionery products. The reverse can be observed in the other member countries of Mercosur, which export the majority of peanuts they produce (Governo do Estado de São Paulo, 1997).

The national production of peanuts, 90% of which is cultivated in the state of São Paulo, has been running at about 150,000 tons/year, with an annual economic value of 50 million U.S. dollars. This production is concentrated in medium-sized properties and no longer in small holdings, with some planting up to 1,000 hectares (Governo do Estado de São Paulo, 1999).

Inadequate procedures, especially during harvesting, drying and storage, resulting in the contamination of the food, are common. Further, technologies such as artificial drying of peanuts in the shell, which could lead to a reduction in the levels of aflatoxin, are still unavailable to the producers at reasonable costs. Some large companies that process peanuts attempt to control mycotoxins, and are able to place safe products on the market. However, smaller producers, for lack of knowledge, lack of capital, or due to the inefficiency of the public inspection process, do not carry out these controls, and sometimes even use products rejected by larger processing companies (Fonseca, 1994).

Although rural extension programs in Brazil have improved in recent years, they are yet to resolve the problem of aflatoxin in peanuts. The "Better Quality for Peanuts Program", financed by ABIA (Brazilian Association of Food Industries), ABEA (Brazilian Association of Peanut Exporting Companies), and peanut-producing cooperatives, with the participation of the University of São Paulo, aimed at reducing the levels of contamination in peanuts by using rural extension programs targeting pre- and post-harvest and storage practices (Fonseca, 1994). After its establishment in 1988, the number of batches rejected by food processors for having exceeded the maximum permitted levels of aflatoxin fell from 40.7 to 3% in 1992. However, in 1994, the number of rejected batches increased to 48% due to an abnormal rain pattern in that year (Fonseca, 1994).

It seems that, in addition to the continuing need for extension activities, the producer sector also needs to invest in adequate technology. However, the national market needs to offer a sufficient incentive for this. Further, there is a lack of punitive regulatory controls, which could induce adoption of safer practices by the private sector. In addition, increased cooperation and development of joint actions between the various elements of the peanut supply chain which aim to combat aflatoxin are nonexistent (Governo do Estado de São Paulo, 1999).

The Cold Chain

The Cold Chain and Food Safety

All foods lose quality following harvest, slaughter, and processing. The speed of this loss depends on the nature of the food, its composition, formulation, type of packaging, storage, and distribution conditions. The loss of quality is a result of physical, chemical, and microbiological changes. For the majority of foods, refrigeration is not an efficient method of preservation but only retards deterioration by reducing the speed of microbial growth. For refrigerated foods to have desirable shelf life, there must be adequate control of pathogenic microorganisms and all the deteriorative microflora capable of multiplying at refrigeration temperatures. The pathogens representing potential hazards for these products should be given special attention in the phases of product development, storage, and distribution. Potential microbiological risks must be recognized and actions for their control or elimination taken, such as those suggested in the Good Manufacturing Practices (GMP) norms for refrigerated products, and the Hazard Analysis and Critical Control Points system (HACCP), which should be incorporated in refrigeration procedures (Moberg, 1989).

Refrigerated foods include meats, chicken, fresh fish, pasteurized milk, meat products (whole and pre-sliced), and fresh pasta. In addition to traditional foods, Brazil, like most other countries, has increased consumption of processed foods. These foods are either ready to serve or require minimal additional processing, with their quality often maintained by refrigeration. Temperature abuse of refrigerated foods during processing, distribution, and sale, or in the hands of the final consumer, can allow for rapid growth of pathogenic microorganisms, compromising the safety of the product.

Brazil currently has the technology and services to implement an efficient cold chain (Neves Filho, 1999). However, it still wrestles with quantitative and qualitative deficiencies with respect to storage, transport, and distribution of refrigerated and frozen products. There is a deficient infrastructure in the country for storage under artificial atmosphere, with only 57 operations and 103 units available for public use. Of these, 14% belongs to the government and 86% is privately owned (Pinazza and Alimandro, 1999). Table 5.2 shows that the country has the capacity to store 3,245.60 tons/day of refrigerated and frozen products. Table 5.2 also shows that warehouses are concentrated in the southern and southeastern regions of the country, while the northern region has only 2.7% and the northeastern region 5.4% of the total refrigerated storage capacity.

In Brazil, 70% of all agricultural products are transported by truck, with the remaining 30% transferred by railroad, waterways, or sea. A small amount is transported by air freight. Pinazza and Alimandro (1999)

Table 5.2 Capacity for Refrigerated Storage for Public Use per Region and in Brazil—1998

Region	*Tons/Day*	*%*
South	1,100.50	33.5
Southeast	1,822.20	56.1
Central west	59.90	1.8
Northeast	176.00	5.4
North	87.00	2.7
Total	3,245.60	100.0

Source: Adapted from Pinazza and Alimandro (1999).

considered refrigerated transportation in Brazil to be inadequate. The authors added that in 1994 less than one quarter of the Brazilian fleet of one million trucks was in good operating condition.

It can be seen in Table 5.3 that the majority of homes in the main metropolitan centers of Brazil have refrigerators, with the exception of the lowest income group. Of those earning up to two minimum salaries and those earning between two and three minimum salaries, 65.8 and 79.3%, respectively have a refrigerator in the home (Instituto Brasileiro de Geografia e Estatística, 1999). It is assumed that in rural areas this proportion is much lower, especially for those with smaller incomes.

Table 5.3 Number of Homes with Refrigerators per Class of Monthly Income—Brazil (1996)

Classes of monthly home income	*Number of homes*	*Number of refrigerators*	*% Refrigerators per home**
Up to2	1,296,406	905,668	69.8
More than 2 to 3	1,040,604	825,223	79.3
More than 3 to 5	1,895,480	1,675,294	88.3
More than 5 to 6	886,445	833,237	93.9
More than 6 to 8	1,380,715	1,307,806	94.7
More than 8 to 10	970,772	939,029	96.7
More than 10 to 15	1,670,694	1,623,133	97.1
More than 15 to 20	965,711	937,702	97.0
More than 20 to 30	947,182	911,352	96.2
More than 30	1,457,297	1,436,425	98.5

Source: Obtained from the Brazilian Institute of Geography and Statistics (IBGE), Survey of Family Income (POF), IBGE (1999).

* Approximate value, since one home may have more than one refrigerator

In Brazil, legislation establishing minimum conditions for the cooling and conservation of food products is aimed at acceptable quality for the consumer. The interministerial decree (Ministries of Health and Agriculture, CISA No. 10, July 31, 1984) establishes a maximum recommended temperature for refrigerated foods of 10°C, and for frozen foods −8°C. It also establishes that for pre-packaged foods, the preservation conditions for transport, marketing, and consumption should be printed on the package. Decree No. CVS 15, July 11, 1991 establishes required temperatures for the transport and delivery of highly perishable foods. Decree No. 304, April 22, 1996, of the Ministries of Agriculture, Supply and Agricultural Reform, established a maximum temperature of 7°C for the delivery of all beef, pork, and offal products marketed.

Although no thorough analysis of the technical problems in the cold chain in Brazil exists, the limited available research paints a worrisome picture. One study carried out by Arruda et al. (1996) considered the adequacy of the conditions for delivery of perishable foods to food service operations, showing that of 369 foods tested, only 28% were delivered according to the temperatures stipulated in CVS 15 (−15 to −18°C for frozen food, 6 to 10°C for cooled foods, and 4 to 6°C for refrigerated goods). In a similar study carried out by Silva and Goes (1999) in the city of Salvador (BA), only 12% of food products were delivered at temperatures established by law (Decree No. 15 CVS).

In November 1996, the National Institute of Metrology (INMETRO)[3] monitored the temperatures of closed horizontal freezers (with lids) and open ones (without lids) used in retail stores. The survey was carried out in six capitals: Porto Alegre, Rio de Janeiro, Brasília, São Paulo, Salvador, and Recife. In each capital, three supermarket chains were chosen, and the measurements were carried out in three shops at each chain by technicians using previously calibrated thermometers. The results showed that the temperatures in 61% of the freezers analyzed were not in accordance with the recommendations of the Ministries of Health and Agriculture. Violations were as follows: Brasília 52%, Porto Alegre 79%, Recife 32%, Rio de Janeiro 62%, and São Paulo 52%. The closed freezers showed much better performance, with only 21% of violations, whereas violations for the open freezers were 76%. In addition, temperatures indicated by the thermometers installed in the freezers were considerably different from those of the calibrated thermometers. According to this study, the supermarkets stated not being aware of the law (Instituto Nacional de Metrologia, Normalização e Qualidade Industrial, 1999a).

---

[3]INMETRO is not an organization that can take punitive actions, such as fines or closures, against food establishments. One of the main objectives of the institute in carrying out tests is to educate the consumer.

Monitoring refrigerated equipment is part of the activity of the Ministries of Health, Agriculture and Supply. However, there is a lack of an adequate infrastructure to carry out these inspections. According to Rissatto (1999), the agents frequently do not even have thermometers to perform this task, as observed in the city of Campinas, an important commercial center.

Considering these data, it is apparent that there are serious problems related to the structure and control of the cold chain in Brazil. Errors highlighted in the increasing distribution and marketing of products which depend on the cold chain for quality and safety maintenance (frozen, chilled, and refrigerated) have given rise to increasing risks for the population.

Refrigerated Milk

The total production of milk in Brazil has increased in recent years. Table 5.4 shows that this value grew from 14,484 million liters in 1990 to 20,352 in 1997. However, alongside this increase there was a rise in the unofficial production of milk. This is defined as milk which is not under the control of the official services for sanitary inspection, and for which no taxes are paid. It has been estimated that in 1990, 25.8% of the total milk production was unofficially produced, with this percentage rising to 48.1% by 1997 (Nascimento and Silva, 1999). This fact demonstrates the inefficiency of the public inspection system for sanitary control of the product. It also suggests that an important number of producers operate with a lack of adequate technology and knowledge, considering that the unofficial

Table 5.4 Milk Production in Brazil, 1990 to 1997 (millions of liters)

Year	*Total Production*	*Formal Production Inspected—FIS*	**Unofficial Production*	*% Production Unofficial*
1990	14,484	10,747	3,737	25.8
1991	15,079	10,413	4,666	30.9
1992	15,784	10,700	4,084	25.8
1993	15,591	9,146	6,445	41.3
1994	15,784	9,441	6,343	40.2
1995	17,694	10,577	7,117	40.2
1996	19,021	11,366	7,655	40.2
1997	20,352	10,557	9,795	48.1

Source: Adapted from Nascimento and Silva (1999).
* Unofficial production is not inspected by the Federal Service of Inspection (FIS) of the Ministry of Agriculture and Supply (MAS)

market includes non-specialized producers, who have no resources to modernize or increase their production scale or margin of profit (Nascimento and Silva, 1999).

The consumption of raw milk in the country is very small, representing 3.9% of the milk consumed in metropolitan regions (Instituto Brasileiro de Geografia e Estatística, 1999), and perhaps slightly more in rural regions, where more recent data are unavailable. It is higher in the metropolitan areas, such as Porto Alegre, Fortaleza, and Belo Horizonte (Instituto Brasileiro de Geografia e Estatística, 1999). Raw milk and its derivatives can transmit diseases such as tuberculosis, diphtheria, brucelosis, and a series of other agents causing gastroenteritis, such as *Salmonella, Staphylococcus aureus,* and others.

Milk should be cooled immediately after milking and maintained below 5°C until consumption. Faults in the refrigeration of pasteurized milk anywhere along the production chain can result in the rapid growth of bacterial flora. However, refrigeration of milk after milking and during transportation to the food processor is not mandatory in Brazil. Indeed, the first insulated tanks were only acquired in the 1980s (Rentero, 1997). Milk with a high initial level of contamination prior to pasteurization can present an important residual flora, leading to problems of product safety. Important microorganisms such as *S. aureus* are destroyed by pasteurization. However, its enterotoxin, capable of producing a severe attack of gastroenteritis, is heat stable and survives in pasteurized foods.

In 1997, INMETRO conducted a survey of the quality of pasteurized milk and showed that of a total of 19 brands marketed in 4 states (Minas Gerais, Rio Grande do Sul, Rio de Janeiro, and São Paulo), 12 brands were deemed to pose a health risk (Instituto Nacional de Metrologia, Normalização e Qualidade Industrial, 1999b). In February 1999, this same institute conducted further analyses of pasteurized milk sold in various regions of the country. Of 26 brands evaluated, 6 were considered unsuitable[4] and 8 unacceptable[5] for human consumption. In about 40% of the sales outlets, the milk was not stored at an adequate temperature (up to 10°C). Thus, even if the milk arrived at the sales outlet with acceptable quality, the microbial load could increase significantly during storage.

The need to increase productivity and therefore competitiveness of the Brazilian milk sector, especially due to the entrance of products from

---

[4]Unsuitable for consumption: number of bacteria × 100 above the limit fixed by decree No. 451, July 2, 1998 of the Secretariat for the Vigilance of Sanitation. The product presents health risks to those who consume it.

[5]Unacceptable for consumption: number of bacteria × 10 above the limit fixed by decree No. 451, July 2, 1998 of the Secretariat for the Vigilance of Sanitation. The product presents health risks to the consumer, the risk varying according to the type of bacteria found.

Argentina and Uruguay with the advent of Mercosur, has resulted in several structural modifications over the last few years. The National Program for the Improvement of Milk Quality (PNQL) instituted by the Ministry of Agriculture and Supply, has, among other measures, fixed a target for the year 2002 for implementing refrigeration of raw milk (maximum of 7°C, 3 hours after milking), and a maximum of 10°C as the temperature on delivery to a food processing facility (anonymous, 1999).

The problem of lack of capital for milk refrigeration equipment in certain production regions has been overcome in various ways. For example, fourteen small producers in the region of Alta Mogiana united to buy a community cooling tank with a capacity of 2500 liters, thus making the process of quality guarantee a reality (Koga, 1999). In addition, smaller tanks have become available on the market with a capacity of 150 liters (Koga, 1999). Some small producers implemented community refrigeration tanks with public support by the concession of physical space or technological extension activities. Large processing companies, worried about the yield and quality of their products, are providing incentives for the use of cooling tanks by their suppliers.

Food Poisoning Outbreaks

Occurrence

The epidemiology of foodborne diseases deals with the occurrence of infections and intoxications transmitted by foods, and has become an essential tool in planning strategies to control and prevent such diseases. However, there is a lack of official data describing the true hygienic and sanitary conditions of the food supplied to the Brazilian population. Sparse official data and some accounts of foodborne illness outbreaks in specific regions and involving specific foods are available as a result of individual research at different research institutes and universities.

For the purposes of this chapter, we will analyze the official food poisoning database of the Ministry of Health - SUS (National Health System) and Center for Vigilance of Sanitation/DITEP, Secretariat of Health, State of São Paulo. In a survey carried out by the Ministry of Health over the period of 1986 to 1997 (Ministério da Saúde, 1999), summarized in Table 5.5, it was found that there were 514,150 hospitalizations due to food poisoning (using the International Classification of Diseases [ICD-9] classification system). In the same period, there were 54,154 hospitalizations in the state of São Paulo, corresponding to 10.53% of the national total. Deaths due to food poisoning in Brazil from 1986 to 1995 amounted to 1119 cases (ICD-9). In the same period, 112 deaths occurred in the state of São Paulo, corresponding to 10% of the total.

Table 5.5 Hospital Mortality and Sickness from Food Poisoning (1986–1997) ICD-9 (International Classification of Diseases)—Brazil and the State of São Paulo

	HOSPITALIZATIONS		*DEATHS*	
Year	Brazil	São Paulo	Brazil	São Paulo
1986	42,033	4,974	156	29
1987	49,510	6,634	166	20
1988	52,414	6,468	143	16
1989	51,320	6,108	151	09
1990	49,169	5,380	110	06
1991	51,870	4,918	82	05
1992	44,264	4,717	82	07
1993	43,247	4,615	92	07
1994	42,081	4,810	68	07
1995	33,919	2,442	69	06
*1996	27,735	1,635	—	—
*1997	26,588	1,453	—	—
TOTAL	514,150	54,154	1,119	112

Source: Adapted from the Ministry of Health, SUS (National Health System).
* Data for deaths in 1996 and 1997 used a new configuration, ICD-10, and cannot therefore be compared since the method for disease notification was different.

It must be mentioned that the data for illnesses refer only to food poisoning outbreaks notified to hospitals, and therefore does not provide the whole picture of the total number of cases occurring in Brazil and in the state of São Paulo. Cases attended in outlying public health units, in private clinics, and those treated at home are not included. The notification of ICD-9 cases is only mandatory for public hospitals. Other sectors of the healthcare industry are not required to report these cases[6] (Fundação Nacional da Saúde, 1999).

In 1994, the Vigilance System for Food Borne Diseases (VETA) was created under the coordination of the Technical Division of Products (DITEP) of the Center for the Vigilance of Sanitation (CVS) of the State Health Secretariat. This system is fundamentally important for the analysis and relevance of outbreak data for foodborne diseases (FBD) noted and investigated by the regional health departments. However, this program is

[6]It must be added that within the large country of Brazil, considerable socio-economic, cultural, geographic, and climatic variations are observed. The distribution of foodborne diseases is probably also varied. In poorer regions, such as the north and northeast, the problem of malnutrition undoubtedly intensifies the problem.

a recent development, and is restricted to the state of São Paulo, one of the most developed areas of the country. Even so, only those cases notified and characterized by official agents of the Vigilance of Sanitation Board were considered. As can be seen from Table 5.6, the number of cases is very small, and the need to improve reporting becomes clear, since there is no doubt that many other outbreaks of FBD have not been included.

With respect to the types of FBD most frequently diagnosed, undoubtedly those related to deficiencies in the hygienic and sanitary conditions are most common. These are associated with ready-to-eat foods, as can be seen from the outbreaks referred to in 1995 and 1996 in the state of São Paulo (Governo do Estado de São Paulo, 1998a). Of the most frequently found FBD, salmonellosis predominates, followed by poisoning by staphylococcal enterotoxins, and more recently by enteropathogenic strains of *Escherichia coli* and others. Outbreaks caused by *Bacilus cereus* (Midura et al., 1970) and *Clostridium perfringens* (Varnam and Evans, 1991) also appear, with these microorganisms usually linked to under-processing rather than to deficiencies in hygienic or sanitary conditions (Table 5.6).

An analysis of the data also reveals a high percentage of outbreaks of unknown origin, both with respect to the source of contamination and the type of organism responsible. No reports on outbreaks of diarrhea caused by viruses have been discovered to date. Such FBD outbreaks have occurred in developed countries, as in the case of the Norwalk type gastroenteritis and hepatitis A (Varnam and Evans, 1991).

Another important aspect of FBD occurrence is related to contamination spread by water. According to data provided by the Division of Water

Table 5.6 Foodborne Disease Outbreaks in the State of São Paulo According to Etiological Agent, 1995/1996

Etiological Agent	*1995*	*1996*	*Total*	%
Salmonella sp	14	11	25	34.2
Salmonella enteritides	7	4	11	15.0
Staphylococcus aureus	5	2	7	9.5
Escherichia coli	1	0	1	1.3
Clostridium perfringens	1	0	1	1.3
Bacilus cereus	1	1	2	2.7
Shigella flexneri	1	0	1	1.3
Indeterminate	14	11	25	34.2
Total	44	29	73	100.0

Source: State Health Secretariat, Center for the Vigilance of Sanitation, Technical Division for Products Related to Health (DITEP).

Borne Diseases of the Center of Epidemiological Vigilance, about 13% of the outbreaks of diarrhea have water as a common source (Governo do Estado de São Paulo, 1998b). These data are important due to the widespread use of water throughout food processing and preparation, undoubtedly serving as a means of spreading contamination. In the survey carried out in the state of São Paulo, restaurants were found to be responsible for approximately 22% of the total outbreaks, followed by bars, drug stores, and similar establishments, which were responsible for approximately 20% of outbreaks. Preliminary studies indicate that the incidence of FBD can be associated with the production of food well before consumption, to contaminated raw materials, and to cross-contamination. In the two years studied, outbreaks in residences represented only 15.1% of the total cases, those of undetermined cause 2.7%, and the rest (82.2%) were in food service establishments (Governo do Estado de São Paulo, 1998a). Given these data, we decided to analyze the case of restaurants in more detail.

Food Safety in the Food Service Industry

In Brazilian urban centers about 25% of meals are prepared outside the home. It has been calculated that some 41 million meals were served per day in 1996. The food service market has been growing intensely, with a growth rate of about 132% between 1994 and 1998. Of the total income of 22.5 billion U.S. dollars in 1995, traditional restaurants earned 9%, fast food outlets, lunch and snack bars 29%, bars 27%, and supermarkets, general stores and bakeries 27%, with the rest attributed to other types of establishments (Associação Brasileira da Indústria de Alimentação, 1999b).

Factors related to raw materials, hygienic conditions, inadequate practices, operations, utensils and equipment, and water supply are critical concerns for food service firms (Lima, 1998; Livera et al., 1996; Organização Panamericana da Saúde and Organização Mundial da Saúde, 1998; Rêgo, 1995; Silva Jr., 1995). One can explain these related FBD occurrences as being partly a result of employees' lack of experience and qualifications. In Brazil, initial research has demonstrated the problem of human resources, mainly at the level of those workers actually preparing the food, a large proportion of whom are semi-illiterate (Lanzillotti, 1994). There is also rapid worker turnover, with workers frequently executing a great variety of activities, resulting in lack of professionalism in the food service sector, contributing to the difficulty in implementing technological innovations (Proença, 1997).

Larger firms, who apply technical criteria of quality and safety, more easily solve the problem of selecting reliable suppliers of raw materials. However, in general this appears to be a widespread and important problem, leading Proença (1997) to recommend that suppliers develop the culture of service quality, including establishing the confidence of restaurant customers.

With the objective of guaranteeing food safety for the population in 1993, by way of decree No. 1, 428, the Ministry of Health made HACCP mandatory in all establishments involved in food preparation and distribution. Many large food service firms have already implemented this system. However, many small and medium-sized food establishments do not even know of this requirement, or of the GMP norms (Buchweitz and Salay, 2000). This fact reiterates the need for improved public communication about food laws. Although cost was the second most common reason given by food services to explain their non-adoption of HACCP, it has been shown that the cost of implementing the system only represents 0.09 to 0.24% of their total annual sales.

In general, food services are the type of food establishment most frequently inspected by the Vigilance of Sanitation inspectors of the Ministry of Health. However, in practice, inspection of an establishment usually occurs either at the time it is registered by the appropriate ministry or as a result of a consumer complaint. The practice of inspection as a preventative and educational measure is not common in Brazil (Rissatto, 1999; Salay and Caswell, 1998).

The main objective of a recently launched joint program by CNI (National Confederation of Industries), SENAI (National Service for Industrial Training), and SEBRAE (Support Service for Small Enterprises) is to train the industry about the HACCP system and support its implementation. At the national launch of this program, it was stated that the goal was to train 500 consultants and 1200 industrial technicians by the year 2001, then train 3500 professionals working within the area (anonymous, 1999).

Lessons Learned and Future Research Needs

Analysis of data on the occurrence of mycotoxins showed that in Brazil, peanuts and corn have the highest levels of contamination with aflatoxins and fumonisins, respectively. With respect to other grains, coffee, milk, and fruit juices, the number of samples analyzed was either too small or originated from only a single region of Brazil, suggesting that the current data is insufficient to evaluate the full extent of the problem. Thus, additional investigations of these products are recommended.

The persistent contamination of peanuts in Brazil appears to be connected to a lack of incentives within the domestic market for firms to adopt safer procedures. However, other factors that explain this problem include an inefficient public control system, lack of adequate technology, need for greater frequency of extension activities, and the use of integrated actions to combat the problem throughout the production chain.

It was also shown that the cold chain presented quantitative and qualitative limitations in Brazil. With respect to the safety of milk, the role of the public sector is extremely precarious, with about 48.1% of the national production not passing through any kind of inspection or control. Also, there is still no legislation requiring producers to cool the product after milking. An important part of the dairy sector continues to have inadequate specialized technology and workers. However, technology for cooling on the farm is available, and economically viable modules, applicable even for small producers, are being marketed in Brazil.

With respect to foodborne illness outbreaks, it was observed that early initiatives to discover the real situation in Brazil are incomplete and do not reflect the true national situation. Specifically, in the case of food service industries, a growing market, increased frequency of inspections by the public sector have not been shown to result in greater safety for the population. With the exception of a few large enterprises, these establishments, in addition to confronting important problems with their human resources, are not aware of current food safety legislation, experience difficulties with food suppliers, and conduct inadequate hygienic and sanitary practices.

A common fact in the three cases studied is inefficiency of the public sector in promoting safe foods. It can be assumed that in developing countries financial difficulties contribute to the problem of lack of governmental food safety controls, especially for foods consumed in the domestic market, as in the cases analyzed. It is therefore accepted that the models of government action should be redefined for greater efficiency. To overcome operational problems in the public sector, the Brazilian government recently formed an national agency within the Ministry of Health designed to have greater social control and reduced political interference, with its own financial resources arising from, among other sources, inspection fees imposed on food firms (Presidência da República, 1999). The results of this initiative are yet to be demonstrated, but the concern remains about how municipalities which, for example, do not even have thermometers to monitor the cold chain, could benefit from this new system.

With respect to the private sector, it was shown that in the cases analyzed, the firms confront problems typical of developing countries such as insufficient training of employees, and sometimes lack of available technology and capital for investment. Sectors of Brazilian agribusiness operate unofficially on a substantial scale, as is the case of the milk industry. However, it appears that the number of firms in Brazil concerned about food safety is increasing, as is the number of firms that aim to increase and/or guarantee their business by promoting safety, according to reports of private actions aimed at better controlling peanuts production, the milk refrigeration, and in implementation of HACCP. It is therefore up to private companies to take this opportunity to invest in this market, which will likely expand in the near future.

Finally, the results of the cases analyzed here indicate that the Brazilian population is still being submitted to significant health risks arising from foods, leaving no room for doubt about the need for interdisciplinary studies in this area. Considering the few studies of this subject in the country, innumerable research suggestions could include problems and incentives for firms to adopt adequate technology for the control of fumonisins in corn, an economic and technical analysis of refrigeration in the producer sector for beef, and standardization and improvement of the surveillance of foodborne diseases.

References

ABIA—Associação Brasileira da Indústria de Alimentação, *Balanço da Indústria da Alimentação em 98,* on line document, (http://www.abia.org.br), 1999a.

ABIA, *O Mercado de Food Service no Brasil,* Departamento Econômico, São Paulo, 1999b.

Anonymous, Indústria de Alimentos terão controle de risco, *Gazeta Mercantil,* A 7, 1999.

Anonymous, Leite: Proposta de Modernização, *Balde Branco,* 28, 414, 1999.

Araújo, A.C.P., Telles, D.L., Almeida, M.G., Nascimento, E.S., Determinación de aflatoxinas en cacahuetes comercializados en Recife-PE, Brasil, *Alimentaria,* 249, 35,1994.

Arruda, G.A., Popolim, W.D., Fujino, H., Leite, C.L., Ribeiro, L.C., Avaliação das condições de entrega de gêneros perecíveis em unidade de alimentação e nutrição, através do método de Análise de Perigos em Pontos Críticos de Controle (APPCC), *Higiene Alimentar,* 10, 44, 1996.

Badiale-Furlong, E., Souza-Soares, L.A., Dadalt, G.M., Micotoxinas em grãos destinados ao consumo humano no Rio Grande do Sul, in *IX Encontro Nacional de Micotoxinas,* Scussel, V.M. Ed., Florianópolis, 1998, 113.

Baldissera, M.A., Ahmad, S.H.E., Pranke, P.H.L., Heinrichs, C.M., Zanandrea, S., Santurio, J.M., Incidência de aflatoxinas em amendoim e farinha de milho, in *VII Encontro Nacional de Micotoxinas,* São Paulo, 1992, 12.

Brigido, B.M., Badolato, M.I.C., Freitas, V.P.S., Contaminação de amendoim e seus produtos comercializados na região de Campinas—SP, por aflatoxinas durante o ano de 1994, *Rev. Inst. Adolfo Lutz,* 55, 85, 1995.

Buchweitz, M.R.D., Salay, E., Analysis of implementation and costs of HACCP system in foodservice industries in the county of Campinas, Brazil, in *The Economics of HACCP: New Studies of Costs and Benefits,* Unnevehr, Ed., Eagan Press, St. Paul, 2000, ch. 20.

Burroughs, L.F., Stability of patulin to sulfur dioxide and to yeast fermentation, *J. AOAC Internat.,* 60, 100, 1977.

Camargos, S.M., Soares, L.M.V., Sawasaki, E., Bortoletto, N., Gumerato, H.M., Incidência de fumonisinas B_1 e B_2 em 35 variedades de milho cultivadas na estação experimental do Instituto Agronômico em Votuporanga, Estado de São Paulo, na safra de 94/95, in *IX Encontro Nacional de Micotoxinas,* Scussel, V.M., Ed., Florianópolis, 1998, 125.

Castro, M.F.P.M., Soares, L.M.V., Furlani, R.P.Z., Mycoflora, aflatoxigenic species and mycotoxins in freshly harvested corn (*Zea mays* L.): a preliminary study, *Rev. Microbiol.,* 26, 289, 1995.

Christensen, C.M. and Kaufman, H.H., *Grain Storage: The Role of Fungi in Quality Loss,* University of Minnesota Press, Minneapolis, 1969, 3.

Corrêa, B., Galhardo, M., Sabino, M., Costa, E.O., Guimarães, E.O.C.F., and Israel, W.M., Distribution of moulds and aflatoxins in dairy cattle feed and raw milk, in *IX International IUPAC Symposium on Mycotoxins and Phycotoxins,* Rome, 1996, 291.

Costa, L.L.F. and Scussel, V.M., Micotoxinas em feijão (*Phaseolus vulgaris* L.) comercializado no Estado de Santa Catarina, in *IX Encontro Nacional de Micotoxinas,* Scussel, V.M., Ed., Florianópolis, 1998a, 119.

Costa, L.L.F. and Scussel, V.M., Ocorrência de micotoxinas em produtos alimentíceos comercializados no município de Florianópolis—Santa Catarina, in *IX Encontro Nacional de Micotoxinas,* Scussel, V.M., Ed., Florianópolis, 1998b, 130.

Ellis, W.O., Smith, J.P., Simpson, B.K., and Oldham, J.H., Aflatoxins in food: occurrence, biosynthesis, effects on organisms, detection, and methods of control, *Crit. Rev. Food Sci. Nutr.,* 30, 403, 1991.

Engel, G. and Teuber, M., Patulin and other small lactones, in *Mycotoxins—Production, Isolation, Separation and Purification,* Betina, V., Ed., Elsevier, Amsterdam, 1984, 291.

Fonseca, H., Calori-Domingues, M.A., Glória, E.M., Zambello, I.V., Segatti-Piedade, F., Ocorrência de aflatoxinas em amendoim no Estado de São Paulo nos anos de 1990 a 1996, in *IX Encontro Nacional de Micotoxinas,* Scussel, V.M., Ed., Florianópolis, 1998, 118.

Fonseca, H., Contribuição ao estudo da ocorrência de aflatoxina em tortas, farelos e farinhas de amendoim (*Arachis hypogaea* L.) no Estado de São Paulo, *Anais ESALQ,* 25, 47, 1968.

Fonseca, H., Program for the Prevention of Peanut Contamination with Aflatoxin, paper presented at *Second Asian Conference on Food Safety,* Bangkok, 1994.

Fraga, M.E., Direito, G.M., Santana, D.M.N., Barros, G.C., and Rosa, C.A.R., Determinação de aflatoxina B_1, M_1 e AFRo em ovos por cromatografia de camada delgada, in *Anais do I Congresso Latino Americano de Micotoxicologia* e *VIII Encontro Nacional de Micotoxinas,* Rio de Janeiro, 1994, 107.

Fundação Nacional da Saúde, *Vigilância Epidemiológica e Informação,* on-line document, http:// www.fns.gov.br, 1999.

Furlani, R.P.Z., Oliveira, P.L.C., and Soares, L.M.V., Incidência de ocratoxina A em café proveniente de várias regiões produtoras brasileiras, in *IX Encontro Nacional de Micotoxinas,* Scussel, V.M., Ed., Florianópolis, 1998, 117.

Furlani, R.P.Z., and Soares, L.M.V., Survey of aflatoxins, ochratoxin A, zearalenone, and stigmatocystin in tree nuts commercialized in Campinas, São Paulo, *Ciênc. Tecnol. Aliment.*, 16, 238, 1996.

Furlong, E.B., Soares, L.M.V., Lasca, C.C., and Kohara, E.Y., Mycotoxins and fungi in wheat stored in elevators in the state of Rio Grande do Sul, Brazil, *Food Add. Contam.*, 12, 683, 1995.

Governo do Estado de São Paulo, Secretaria de Estado da Saúde—Centro de Vigilância Sanitária—DITEP, *Análise dos Surtos de Enfermidades Transmitidas por Alimentos,* São Paulo, 1998a.

Governo do Estado de São Paulo, Secretaria de Estado da Saúde, Divisão de Doenças de Transmissão Hídrica do Centro de Vigilância Epidemiológica Prof. Alexandre Vranjac, CIP, São Paulo, 1998b.

Governo do Estado de São Paulo, Secretaria de Agricultura e Abastecimento—CATI, Comissão Técnica de Oleaginosas, *Amendoim Produção em São Paulo e Implicações no Mercosul,* Campinas, SP, Brasil, 1997.

Governo do Estado de São Paulo, Secretaria de Agricultura e Abastecimento—CATI, Comissão Técnica de Oleaginosas, *Oleaginosas no Estado de São Paulo Análise e Diagnóstico,* Campinas, 1999.

Hirooka, E.Y., Yamaguchi, M.M., Aoyama, S., Sugiura, Y., Ueno, Y., The natural occurrence of fumonisins in Brazilian corn kernels, *Food Add. Contam.,* 13, 173, 1996.

INMETRO—Instituto Nacional de Metrologia, Normalização e Qualidade Industrial, *Freezers em Supermercado,* on line document, http://www. inmetro.gov.br, 1999a.

INMETRO—Instituto Nacional de Metrologia, Normalização e Qualidade Industrial, *Leite tipo "B", tipo "C", UHT e queijo minas frescal e prato,* on-line document, http://www.inmetro.gov.br, 1999b.

Instituto Brasileiro de Geografia e Estatística—IBGE, *Sistema IBGE de Recuperação Automática—SIDRA,* on line document, http://www.ibge.gov.br, 1999.

Koga, C., Refrigeração do leite chega ao pequeno produtor, *Balde Branco,* 412, 36, 1999.

Kuiper-Goodman, T., Scott, P.M., Watanabe, H., Risk assessment of the mycotoxin zearalenone, *Regul. Toxicol. Pharmacol.,* 7, 253, 1989.

Lanzillotti, H.S., Gerência de Recursos Humanos em Alimentação Institucional, *Revista de Nutrição da PUCCAMP,* 7, 89, 1994.

Lavinas, L., Manão, D., Garcia, E.H., Amaral, M.R., Acessibilidade alimentar e estabilização Econômica do Brasil no anos 90, *Revista Nova Economia,* 8, 59, 1998.

Lazzari, F.A., Ocorrência de toxinas de *Fusarium* em milho no Paraná, in *Anais do I Congresso Latino Americano de Micotoxicologia* e *VIII Encontro Nacional de Micotoxinas,* Rio de Janeiro, 1994, 104.

Leoni, L.A.B., Soares, L.M.V., and Castro, M.F.P.P.M., Levantamento da incidência de moniliformina em milho recém colhido produzido no Estado de São Paulo, em 1992, in *Anais do I Congresso Latino Americano de Micotoxicologia* e *VIII Encontro Nacional de Micotoxinas,* Rio de Janeiro, 1994, 109.

Lima, V.L.A.G., Melo, E.A, and Sena, E.N., Condições higiênico-sanitárias de fast-food e restaurantes da região Metropolitana da Cidade do Recife—PE, *Higiene Alimentar,* 12, 50, 1998.

Livera, A.V.S., Santos, A.C.O., Melo, E.A., Rêgo, J.C., and Guerra, N.B., Condições higiênico-sanitárias de segmentos dacadeia alimentar do Estado do Pernambuco, *Higiene Alimentar,* 10, 28, 1996.

Machinski, M. Jr., Soares, L.M.V., Sawasaki, E., Sorti, G.D., Castro, J.L., and Bortoletto, N., Ocorrência de fumonisinas B_1 e B_2 em milho plantado no Estado de São Paulo, in *IX Encontro Nacional de Micotoxinas,* Scussel, V.M. Ed., Florianópolis, 1998, 114.

Machinsky, M. Jr., and Midio, A.F., Incidencia de patulina en jugo de manzana industrializado, *Alimentaria,* 276, 61, 1996.

Midura, T. et al., Outbreak of food poisoning caused by *Bacillus cereus,* Public Health Reports, USA, 1, 45, 1970.

Ministério da Saúde/SIH/SUS/DATASUS, *Morbidade e Mortalidade Hospitalar,* on line document, http://www.saude.gov.br//, 1999.

Ministério da Agricultura e Abastecimento, Portaria No. 230 de 10 de Junho de 1997, *Diário Oficial da União,* Brasília, June 10, 1997.

Moberg, L., Good manufacturing practices for refrigerated foods, *J. Food Protec,* 52, 363, 1989.

Motta, S. and Soares, L.M.V., Survey of tomato products for alternariol, alternariol monomethylether, tenuazonic acid, and cyclopiazonic acid, in *IX International IUPAC Symposium on Mycotoxins and Phycotoxins,* Rome, 1996, 162.

Nascimento, R.C. and Silva, J.M.A., Investimento Agroindustrial e Capital Estrangeiro no Brasil: O Caso do Setor de Laticínios, in *Anais XXXVII Congresso Brasileiro de Economia e Sociologia Rural,* Foz do Iguaçu, 1999.

Navas, S.A., Zorzetto, M.A.P., and Sabino, M., Incidência de aflatoxina M_1 em leite em pó destinado a merenda escolar por cromatografia em camada delgada. Estudo preliminar, in *Anais do I Congresso Latino Americano de Micotoxicologia* e *VIII Encontro Nacional de Micotoxinas,* Rio de Janeiro, 1994, 102.

Neves Filho, L.C., *A temperatura de estocagem e distribuição de alimentos refrigerados,* on-line document, http://www.recrusul.com.br, 1999.

Oliveira, C.A.F., Germano, P.M.L., Bird, C., and Pinto, C.A., Immunochemical assessment of aflatoxin M_1 in milk powder consumed by infants in São Paulo, Brazil, *Food Add. Contam.,* 14, 7, 1997.

Oliveira, M.C.M., Oliveira, E.J.A., and Moraes, I.M., Comparação da incidência de aflatoxinas em amostras de amendoim e produtos derivados, analisados no Departamento de Toxicologia do Lacen/PE nos anos de 1996 e 1997, in *IX Encontro Nacional de Micotoxinas,* Scussel, V.M., Ed., Florianópolis, 1998, 115.

Ono, E.Y.S., Camilo, S.B., Levy, R.M., Ono, C.J., Rossi, C.M.A., Figueira, E.L.Z., Sá, M.C., Oliveira, T.C.R.M., and Hirooka, E.Y., Microbiota fúngica e aflatoxina B_1 em milho recém-colhido no Estado do Paraná, in *IX Encontro Nacional de Micotoxinas,* Scussel, V.M., Ed., Florianópolis, 1998a, 123.

Ono, E.Y.S., Ueno, Y., Hashimoto, E.H., Funo, F.Y., Ono, M.A., Ono, C.J., Oda, P., and Hirooka, E.Y., Microbiota fúngica e fumonisinas em milho do Estado do Paraná, in *IX Encontro Nacional de Micotoxinas,* Scussel, V.M., Ed., Florianópolis, 1998b, 121.

Organização Panamericana da Saúde—OPAS/Organização Mundial da Saúde—OMS, *A Saúde no Brasil,* Brasília, 1998.

Pich, P.H., Nordin, N.S.D., and Noll, I.B., Detecção de aflatoxinas em produtos derivados de milho comercializados na região de Porto Alegre - RS, in *IX Encontro Nacional de Micotoxinas,* Scussel, V.M., Ed., Florianópolis, 1998, 120.

Pinazza, L.A., Alimandro, R., *Reestruturação no Agribusiness Brasileiro,* Abag. Agroanalysis and FGV, Rio de Janeiro, 1999, ch. 15.

Pohland, A.E., Nesheim, S., and Friedman, L., Ochratoxin A: a review, *Pure Appl. Chem.,* 64, 1029, 1992.

Pollock, G.A., Disabatino, C.E., Heimsch, R.C., and Hilbelink, D.R., The subchronic toxicity and teratogenicity of alternariol monomethyl ether produced by *Alternaria solani, Food Chem. Toxicol.,* 20, 899, 1982.

Pozzi, C.R., Corrêa, B., Gambale, W., Paula, C.R., Chacon-Reche, N.O., and Meirelles, M.C.A., Postharvest and stored corn in Brazil: mycoflora interaction, abiotic factors and mycotoxin occurrence, *Food Add. Contam.,* 12, 313, 1995.

Prado, G., Oliveira, M.S., Ferreira, S.O., Corrêa, T.B.S., and Affonso, B.R.R., Ocorrência natural de desoxinivalenol e toxina T-2 em milho pós-colheita, *Ciênc. Tecnol. Aliment.,* 17, 259, 1997.

Presidência da República, Define o Sistema Nacional de Vigilância Sanitária, cria a Agência Nacional de Vigilância Sanitária, e dá outras providências, Lei No. 9, 782 de 26 de Janeiro de 1999, *Diário Oficial da União,* Brasília, January 27, 1999.

Proença, P.C., *Inovação Tecnológica de Alimentação Coletiva,* Insular, Florianópolis, 1997, ch. 4.

Rêgo, J.C., *Influência do treinamento no controle higiênico-sanitário de unidades de alimentação e nutrição,* Tese de Mestrado, Universidade Federal de Pernambuco, Brasil, 1995, 43.

Rentero, N., Quando qualidade significa resfriamento e coleta a granel, *Balde Branco,* 394, 46, 1997.

Rissatto, P.E.V., *Avaliação da operacionalização do serviço de Vigilância Sanitária de Alimentos do município de Campinas—SP, de 1993 a 1996,* Tese de Mestrado, Faculdade de Engenharia de Alimentos, Universidade Estadual de Campinas, São Paulo, Brazil, 1999.

Rosa, M.F.A.P., Santana, D.M.N., Direito, G.M., Sabino, M., and Rosa, C.A.R., An improved method to determinate aflatoxins B_1 and M_1 and aflatoxicol in Brazilian industrializated liver paste, in *IX International IUPAC Symposium on Mycotoxins and Phycotoxins,* Rome, 1996, 288.

Sabino, M. and Rodriguez-Amaya, D.B., Mycotoxin research in Brazil, *Ciência e Cultura,* 45, 359, 1993.

Sabino, M., Zorzetto, L.C.A., Lamardo, M.A.P., Inomata, E.I., Navas, S.A., and Milanez, T.V., A survey of the occurrence of aflatoxins in peanuts and peanuts products in São Paulo State/Brazil in 1994, in *IX International IUPAC Symposium on Mycotoxins and Phycotoxins,* Rome, 1996, 129.

Salay, E. and Caswell, J.A., Developments in Brazilian food safety policy, *Int. Food Agribusiness Rev.,* 1, 167, 1998.

Scott, P.M., Effects of food processing on mycotoxins, *J. Food Protec.,* 47, 489, 1984.

Scott, P.M., Fumonisins, *Int. J. Food Microbiol.,* 18, 257, 1993.

Silva Jr., E.A., *Manual de Controle Higiênico-Sanitário em Alimentos,* Varela, São Paulo, 1995, ch. 6.

Silva, S.B. and Góes, J.A.G., Avaliação das condições de entrega de gêneros alimentícios perecíveis em unidades de alimentação e nutrição da Cidade de Salvador—Bahia, *Higiene Alimentar,* 13, 10, 1999.

Snyder, A.P., Qualitative, quantitative, and technological aspects of the trichothecenes mycotoxins, *J. Food Protec.,* 49, 544, 1986.

Soares, L.M.V. and Furlani, R.P.Z., Micotoxinas em milho de pipoca, *Bol. SBCTA,* 26, 33, 1992.

Soares, L.M.V. and Furlani, R.P.Z., Survey of aflatoxins, ochratoxin A, zearalenone and stigmatocystin in health foods and breakfast cereals commercialized in the city of Campinas, São Paulo, *Ciênc. Tecnol. Aliment.,* 16, 126, 1996a.

Soares, L.M.V. and Furlani, R.P.Z., Survey of mycotoxins in wheat and wheat products in health food stores of the city of Campinas, State of São Paulo, *Rev. Microbiol.,* 27, 41, 1996b.

Souza, L.M., Almeida, J.L., and Andrade, I.A., Ocorrência de Micotoxinas em Produtos, Subprodutos e Derivados de Origem Vegetal: Relatórios Anuais dos Laboratórios Credenciados para Análise de Micotoxinas da Rede Oficial do Ministério da Agricultura e do Abastecimento. Período: 1996 e 1997, in *IX Encontro Nacional de Micotoxinas,* Scussel, V.M., Ed., Florianópolis, 1998, 128.

Stoloff, L., Van Egmond, H.P., and Park, D.L., Rationales for the establishment of limits and regulations for mycotoxins, *Food Add. Contam.,* 8, 213, 1991.

Sylos, C.M., Ocorrência de micotoxinas em alimentos Brasileiros, in *Anais XVI Congresso Brasileiro de Ciência e Tecnologia de Alimentos,* Figueiredo, A.A., Ed., Rio de Janeiro, 1998, 314.

Sylos, C.M., Rodriguez-Amaya, D.B., and Carvalho, P.R.N., Occurrence of aflatoxin M_1 in milk and dairy products commercialized in Campinas, Brazil, *Food Add. Contam.,* 13, 169, 1996.

Sylos, C.M. and Rodriguez-Amaya, D.B., Ocorrência simultânea de ácido ciclopiazônieco aflatoxinas B e G em amendoim e milho, in *Anais do I Congresso Latino Americano de Micotoxicologia* e *VIII Encontro Nacional de Micotoxinas,* Rio de Janeiro, 1994, 83.

Sylos, C.M. and Rodriguez-Amaya, D.B., Incidence of patulin in fruits and fruit juices marketed in Campinas, Brazil, *Food Add. Contam.,* 16, 71, 1999.

Sylos, C.M., Rodriguez-Amaya, D.B., Santurio, J.M., and Baldissera, M.A., Occurrence of aflatoxins and cyclopiazonic acid in Brazilian peanut and corn, in *IX International IUPAC Symposium on Mycotoxins and Phycotoxins,* Rome, 1996, 132.

Tango, J.S., Menezes, T.J.B., and Teixeira, C.G., Levantamento da ocorrência de aflatoxina em sementes de amendoim nas safras das águas e da seca, *Col. ITAL,* 1, 1, 1966.

Taniwaki, M.H. and Van Dender, A.G.F., Occurrence of toxigenic molds in Brazilian cheese, *J. Food Protect.,* 55, 187, 1992.

Vale, V.L., Variação dos níveis de aflatoxina B_1 em milho de pipoca comercializado em Fortaleza, durante 1991, in *VII Encontro Nacional de Micotoxinas,* São Paulo, 1992, 13.

Van Egmond, H.P., Aflatoxin M_1: occurrence, toxicity, regulation, in *Mycotoxins in Dairy Products,* Van Egmond, H.P., Ed., Elsevier Applied Science, London, 1989, 11.

Vargas, E.A., Programa Nacional de Controle de Micotoxinas em Produtos, Subprodutos e Derivados de Origem Vegetal (PNCMV), in *IX Encontro Nacional de Micotoxinas,* Scussel, V.M., Ed., Florianópolis, 1998, 29.

Varnam, A.H. and Evans, M.G., *Foodborne Pathogens, an Illustrated Text,* Moshy Year Book, London, 1991, ch. 15, 18.

Vieira, A.P., Badiale-Furlong, E., Avaliação da incidência de micotoxinas e qualidade de farinhas de trigo comercializadas na zona sul do Rio Grande do Sul, in *Simpósio Latino Americano de Ciência de Alimentos,* Campinas, 1995, 124.

Vieira, A.P., Pinho, B., and Badiale-Furlong, E., Avaliação da qualidade e ocorrência de micotoxinas em farinhas, pães e pizzas comercializadas na zona sul do Rio Grande do Sul, in *IX Encontro Nacional de Micotoxinas,* Scussel, V.M., Ed., Florianópolis, 1998, 112.

World Bank, *World Development Indicators,* on-line document, http://www.worldbank.org, 1999a.

Chapter 6

Product Liability and Food Safety: The Resolution of Food Poisoning Lawsuits

Jean C. Buzby, Paul D. Frenzen, and Barbara Rasco

The U.S. product liability system was designed to alleviate the economic costs that individuals incur due to defective products, including food products contaminated by microbial pathogens. Under current law, consumers who suffer from a foodborne illness can seek monetary compensation for their injuries through the courts. Firms producing these defective products are liable for medical and other costs to ill consumers. Firms may incur other costs such as legal fees, higher insurance costs, and lost market share and profits. Economic theory suggests that foodborne illness litigation is a signal for firms to invest more in food safety, ultimately resulting in a lower incidence of foodborne illness and an increase in general social welfare. However, the vast majority of consumers who experience a foodborne illness do not file a legal claim, or drop or resolve their claim prior to court proceedings, almost always out of public view. Data on legal outcomes for food poisoning lawsuits are scarce, and most available data are for cases that result in jury verdicts. This preliminary study

0-8493-2217-0/01/$0.00+$.50

analyzes a sample of jury verdicts in food poisoning cases in order to assess the economic incentives provided by the legal system for firms to produce safer food.

Introduction

Humans may become ill when they eat foods contaminated by microbial pathogens, which include bacteria, parasites, viruses, and fungi and their toxins. Foodborne illness is relatively common in the United States despite intensive efforts to ensure the safety of food products. Most foodborne infections are not particularly severe and typically involve brief episodes of nausea or diarrhea. However, some foodborne infections cause serious or fatal health problems, and about 2 to 3% of these infections result in chronic illnesses, such as reactive arthritis or Guillain-Barré syndrome (Archer and Kvenberg, 1985). The U.S. Centers for Disease Control and Prevention (CDC) recently estimated that foodborne microbial pathogens are responsible for 76 million annual illnesses in the U.S., resulting in 325,000 hospitalizations and 5000 deaths (Mead et al., 1999). Pathogen-contaminated foods consequently represent an important cause of unintentional injury and death. In fact, contaminated food products are responsible for more deaths each year than the 15,000 consumer products regulated by the U.S. Consumer Product Safety Commission, which were collectively associated with approximately 3700 accidental deaths in 1996 (U.S. Consumer Product Safety Commission, 1998).

Under U.S. product liability law, individuals harmed by unsafe products, including pathogen-contaminated foods, can take legal action to obtain financial compensation for their injuries. Product liability law describes the circumstances under which an individual can recover damages for a defective food item. The laws of individual states govern the nature and extent of compensation that may be awarded for injuries or deaths due to contaminated food products. Product liability litigation makes it possible to shift the economic costs of foodborne illness from consumers to the firms responsible for causing illness. Economic theory also suggests that foodborne illness litigation is a signal for firms to invest more in food safety, ultimately resulting in a lower incidence of foodborne illness and an increase in general social welfare.

This chapter describes a preliminary interdisciplinary investigation of how product liability actually functions in the case of foodborne illness, based on an analysis of U.S. court cases involving injuries due to pathogen-contaminated food products. We focused on the characteristics of cases decided by jury verdicts, including types of foods and pathogens involved, severity of illness, and size of damage awards. The investigators represent three different disciplines: agricultural economics, demography, and law. The agricultural economist provided an economic perspective on food safety, the demographer contributed an understanding of the dynamic processes involved in foodborne illness, and the attorney clarified the intricacies of product liability law. The next sections of this chapter review the incentives for U.S. firms to produce safe food, and the laws governing liability for foodborne illness. Subsequent sections describe the jury verdict data and results of the analysis. The final section summarizes the most important findings and the need for further research.

The Product Liability System

U.S. firms that make or distribute food products have a variety of incentives to reduce microbial pathogen contamination to safe levels. These incentives are provided by market, regulatory, and legal mechanisms (Garber, 1998), and generally take the form of negative incentives, or adverse consequences for firms responsible for making or distributing pathogen-contaminated food. The basic components of this incentive system are:

1. *Market forces:* firms risk losing business reputation, market share, and sales revenue if consumers become concerned about the safety of the firm's food products.
2. *Food safety laws and regulations:* firms that violate federal, state, or local food safety laws or regulations may be subject to various actions imposed by law, the courts, or government agencies, including fines, product recalls, and temporary or permanent plant closures.
3. *Product liability:* persons made ill by contaminated food products may seek financial compensation for their injuries from the responsible party by pursuing legal action under product liability laws, which establish the circumstances under which injured consumers are entitled to financial damages.

The food safety incentive system is complex because market forces, government regulation, and product liability litigation are constantly interacting and evolving in response to the emergence of new foodborne pathogens, scientific and technological advances, changes in consumer food preferences, new laws and regulations, precedent-setting legal decisions, and other factors. For example, the 1996 outbreak of foodborne illness due to *E. coli* O157:H7 contamination of unpasteurized apple juice manufactured by Odwalla, Inc. raised consumer concerns about the safety of fresh juice nationwide and prompted many juice manufacturers to voluntarily begin pasteurizing juice products that were previously unpasteurized. The increasing number of foodborne illness outbreaks due to *E. coli* O157:H7 contamination of unpasteurized juice products also led the U.S. Food and Drug Administration (FDA) to propose new regulations for juice products (Buzby and Crutchfield, 1999). These changes in market forces and government regulations were in addition to the adverse consequences for Odwalla, which included a voluntary product recall costing $12.5 million, a 17% drop in revenue during the first 6 months after the outbreak, a record $1.5 million federal fine for interstate shipment of an adulterated food product, and 21 personal injury lawsuits (Roach, 1999; Munarriz, 1997).

Although all three components of the food safety incentive system affect firms, product liability litigation has received less attention than market forces or government regulation, perhaps because product liability law is complex, and litigation is usually resolved out of public view. Federal and state laws determine when a defendant firm may be legally responsible for injuries or deaths due to pathogen-contaminated food, but liability also depends on the prevailing judicial interpretation of the law, as well as exactly how a food product became contaminated and caused illness. For example, liability currently varies according to the type of food and pathogen involved in causing illness, reflecting previous legal decisions and judicial assumptions about consumer awareness of the safety of particular foods.

Food products may be contaminated by pathogens in many different ways, further complicating the determination of legal liability. Some pathogens live in the intestinal tracts of domestic animals and may contaminate meat or poultry during slaughter. Other pathogens are widespread in the natural environment, and may be introduced into foods at numerous points during food production, processing, or preparation (CAST, 1994). Many pathogen-contaminated foods may not cause human illness unless consumers make errors in food handling, typically by consuming the food when it is raw or undercooked. Although surveys of consumer behavior suggest that food handling errors are common, there is no reliable estimate of the proportion of foodborne illnesses that are solely due to consumers and do not involve liability by firms.

Consumers are likely to encounter considerable obstacles in proving that a specific firm was responsible for a foodborne illness, because of the relatively long incubation periods for most foodborne infections. The symptoms of acute foodborne infections typically do not appear until several days after the consumption of contaminated food, while chronic complications may not appear until several weeks after the initial infection. The difficulty of tracing a foodborne illness back to a specific food source increases as the incubation period lengthens, multiplying the number of other potential sources of contamination and reducing the chances that any suspect food will be available for microbiological testing. Moreover, most individuals made ill by foodborne pathogens never receive a definitive medical diagnosis, either because they did not seek medical care or because their physician did not test for the pathogen that caused the illness (Frenzen et al., 1999).

Firms that make or distribute food products tend to be concerned about the threat of product liability litigation, and almost all firms involved in the food industry have some form of insurance that will apply in the event of a foodborne illness (Clark, 2000).[1] Insurers also participate in product liability litigation when firms have coverage. Litigation is costly and the outcome of a trial is uncertain, so many firms and their insurers prefer to resolve consumer complaints about pathogen-contaminated food products outside the courtroom. In the case of settlements, insurers generally control litigation and determine the amount and timing of payment (Clark, 2000). These settlements are usually kept confidential, making it difficult to gauge the frequency of consumer complaints or lawsuits about contaminated foods. Some lawsuits ultimately result in public trials, but published trial statistics are limited and do not distinguish among subcategories of product liability cases. As a result, very little is known about the basic characteristics of litigation involving foodborne illness.

U.S. Product Liability Law and Foodborne Pathogens

Product liability law is the area of state common law dealing with personal injuries due to defective products. Although the laws regarding foodborne illness vary from state to state, overall there is a fairly consistent body of state law in this category (Clark, 2000). In contrast, there is no uniform or comprehensive federal law on product liability. Under product liability law, products are considered defective if they pose hazards or are found to be of inferior condition or quality (Keeton et al., 1984). The basic concepts of product liability law, the three main legal grounds for bringing a food poisoning lawsuit, and some of the strategies used in product liability trials are described here. We use the term "food poisoning" to refer to injuries caused by pathogen-contaminated food products.

Consumers who file lawsuits to obtain compensation for damages due to a defective product are called *plaintiffs*. Most plaintiffs in food poisoning cases are consumers seeking compensation for personal illness. However, food poisoning lawsuits can also be brought by parents on behalf of dependent children, by guardians on behalf of legally incompetent individuals, or by estates seeking monetary damages in cases of wrongful death. In some cases, a firm may sue other businesses for loss

[1]Bruce Clark is an attorney at Marler-Clark, a law firm in Seattle that has handled many prominent foodborne illness cases.

of reputation, lost profits, or other damages associated with foodborne illness due to mishandling of the firm's products. A plaintiff initiates a lawsuit by filing a court document called a complaint, which describes the injuries and identifies the party, or *defendant,* who is allegedly responsible for the injuries. The defendants in food poisoning cases may include firms that produce, process, distribute, or sell food products. Defendants may occasionally include individual proprietors, employees such as food servers, or even hosts of informal meals or other private events where food was served.

Many food poisoning lawsuits never reach the courtroom. Some cases are dismissed at early stages because the plaintiff did not follow proper court procedures, or because the defendant is not within the jurisdiction of the court. Other cases are settled out of court when firms (or their insurers) agree to pay compensation to consumer plaintiffs in return for dropping their lawsuit before trial. Cases may also be resolved outside court through more formal methods of alternate dispute resolution such as mediation or arbitration. Plaintiffs with strong legal cases are likely to receive early settlement offers, possibly even before filing a lawsuit (Rosenbaum, 1998). Firms (and their insurers) have incentives to settle out of court in order to limit their legal costs, avoid the uncertain outcome of a public trial, and avert the potentially adverse impact of a public trial on the reputation of the firm or its products. National firms and well-known firms are particularly likely to settle cases quickly (Rosenbaum, 1998). Settlement terms are usually kept confidential in order to protect the firm's business reputation and avoid encouraging copycat claims and lawsuits. Some settlements are reported in court records or other accessible sources, but these cases are unlikely to be representative, because plaintiffs' attorneys may selectively divulge only the most favorable settlements in order to boost their professional reputation. The lack of comprehensive public information about settlements makes it difficult to assess the overall frequency or characteristics of food poisoning litigation.

Grounds for Bringing Food Poisoning Lawsuits

State courts recognize three main causes of action, or grounds for bringing a product liability lawsuit: (1) strict liability, (2) negligence, and (3) breach of warranty.

Strict Liability

Strict liability is often cited in food poisoning lawsuits when food is defective or dangerous. The plaintiff must prove that the product was both defective and unreasonably dangerous in order to recover damages, but need not prove that the defendant was negligent (Rasco, 1997). In particular, the plaintiff must show that the defendant's failure to make the product safe was the *proximate cause* of the plaintiff's illness. *Proximate cause* is the legal term for the causal link between a defective product and the resulting injury (Harl, 1997).

Since the courts generally recognize that food products cannot be made completely risk free, strict liability is usually unsuccessful as a cause of action. Food poisoning lawsuits that raise strict liability alone may be dismissed in cases involving naturally occurring pathogens such as *Vibrio vulnificus* in raw oysters. However, manufacturers may be required to warn consumers of the potential hazards associated with such products. A failure to warn consumers of hazards can result in negligence claims (see below), particularly in situations where legally required warnings were not provided. Examples of legally required warnings include restaurant health warnings about the consumption of raw molluscan shellfish, retail store warnings about the consumption of raw milk products and unpasteurized juices, and federally mandated safe handling and cooking labels on raw meat.

Negligence

Negligence can be raised as a cause of action for a food poisoning lawsuit when the defendant failed to exercise reasonable care in producing, marketing, distributing, or selling food. The plaintiff must show that the defendant, through neglect or carelessness, failed to prevent the product from becoming defective and harming the user. It is unlikely that plaintiffs would base a foodborne illness lawsuit solely on a negligence claim (Clark, 2000). The legal theory of *negligence per se* is particularly relevant for food poisoning lawsuits. Under *negligence per se,* it is some evidence that a defendant was negligent if he or she violated a law or regulation that was specifically designed to prevent the type of injuries that the plaintiff suffered (Rasco, 1997).

Claims of negligence can be extended to middlemen such as distributors or retailers if they failed to exercise reasonable care in handling a food product, for example by failing to refrigerate it at the proper temperature. However, retailers are unlikely to be held liable for latent defects in food products unless they also produced it or were clearly expected to test the safety of the product (Harl, 1997). Retailers are generally expected to inspect the food they sell, and may be found negligent if they failed to use a feasible inspection procedure (*American Law of Products Liability,* 1987). Although some courts have ruled otherwise, retailers are usually not expected to open sealed food containers in the course of their inspection (*American Law of Products Liability*, 1987).

Another factor determining liability is the extent to which consumers take precautionary measures when handling food. Under the legal theory of *contributory negligence,* consumers are partially responsible for injuries they suffer due to their failure to exercise reasonable care. For example, the courts do not consider raw pork to be defective and unreasonably dangerous to consumers when it is contaminated by the tapeworm *Trichinae* because most consumers are assumed to be aware of the risk of eating undercooked pork, and know that the risk can be eliminated by thorough cooking (*American Law of Products Liability*, 1987; *Restatement of the Law*, 1965).

Consumers may have no grounds for bringing food poisoning lawsuits if they were aware of the health risks of consuming a specific food prior to consuming it and becoming ill (*American Law of Products Liability*, 1987). For example, a consumer with cirrhosis of the liver who became ill after consuming raw oysters despite being informed about the health hazards of a *V. vulnificus* infection is unlikely to recover damages from either the seafood company that produced the oysters or the restaurant that served them. Defendant firms have sometimes argued that a plaintiff was negligent in consuming a particular food because he or she had a preexisting medical condition that increased his/her susceptibility to foodborne pathogens, but the effectiveness of this argument has not been established.

Breach of Warranty

Breach of warranty is a legal theory drawn from commercial law, and has generally been the most successful cause of action for bringing a food poisoning lawsuit. Under the *Uniform Commercial Code*, a seller incurs obligations called *warranties* by selling a product, and plaintiffs who bought the product can recover damages if it did not conform to warranty and was not fit to eat. Plaintiffs only need to prove that the defendant sold an unwholesome or defective food that caused the plaintiff to become ill.

Two different kinds of warranties known as *express warranties* and *implied warranties* are relevant in the case of food products. An express warranty is an explicit part of the bargain when a product is sold and includes advertisements, pictures, or declarations on the food container or menu, or statements made by sales representatives about the material characteristics of the product. An example of an express warranty is the declaration on a egg carton that the eggs are "*Salmonella*-free." A food seller is liable for a breach of express warranty if the plaintiff contracts salmonellosis because this warranty was not true.

An *implied warranty of merchantability* exists whenever a firm sells the goods that represent the business of the firm. Under an implied warranty, food products must meet the level of quality considered generally acceptable in the marketplace (Rasco, 1997). A food product must also meet prescribed safety standards and be fit for the ordinary purpose for which it is sold, namely human consumption (*The Guide to American Law*, 1984). Fitness for consumption is determined by how a consumer might reasonably expect to use the product (*Gates v. Standard Brands, Inc.*, 1986).

A product must also meet an *implied warranty of fitness*. A seller makes an implied warranty of fitness when he knows the buyer's intended use of a product, and the buyer relies on the seller's judgment or skill in selecting a suitable product. For example, if a retailer informed a consumer that a particular type of raw fish could be used to make sushi and the consumer subsequently became infected by *Anisakis simplex* worms after consuming the sushi, the implied warranty of fitness for the fish would be breached, and the retailer would be liable for damages.

Strategies Used in Product Liability Trials

Court trials of food poisoning lawsuits are conducted in two stages. The first stage is the determination of liability, and the second stage is the assessment of damages if liability is found. The determination of liability depends in part upon whether there is clear evidence of a causal link between the defendant firm and the damages suffered by the consumer plaintiff. The establishment of a causal link between a specific food product and a foodborne illness is often difficult for the reasons noted above, including the number of other potential sources of infection encountered during the incubation period, the limited chances that any suspect food will be found for microbiological testing, and the frequent lack of a definitive medical diagnosis. Plaintiffs must overcome each of these obstacles to establish that a foodborne illness was due to the consumption of a specific food product, and then must show that the actions of the defendant firm caused the food product to become contaminated by pathogens.

Defendant firms can usually defeat negligence claims in court by demonstrating that the firm exercised reasonable care in producing, handling, or distributing the suspect food product, and also obeyed all relevant food safety laws and regulations. The use of state-of-the art food safety procedures provides some evidence that a firm exercised reasonable care, but is not sufficient alone to defeat a negligence claim. Evidence that a firm exercised reasonable care and obeyed relevant laws and regulations is also insufficient to defeat a breach of warranty claim, but may help mitigate any damages that might be assessed against the firm.

Experimental Design Method: U.S. Food Poisoning Cases

We focused exclusively on food poisoning lawsuits that were tried and resulted in jury verdicts because there is no source of representative data on lawsuits that were dropped or settled prior to trial. Information about lawsuits that resulted in jury verdicts was obtained by systematically searching two major jury verdict databases, the Westlaw® Jury Verdicts and Settlement Summaries (West Group, Inc., Eagan, Minnesota) and the LEXIS-NEXIS® Verdicts Library (Reed Elsevier plc, London, England). Both databases consist of descriptive summaries of civil jury verdicts, as well as some out-of-court settlements, prepared by jury verdict reporters, private firms that collect and sell this information to practicing attorneys. Most jury verdict reporters cover a single state or metropolitan area. Only a few of the jury verdict reporters collect every verdict in their covered area. However, the others collect a sample of verdicts without any specific biases in favor of plaintiff decisions or large damage awards. The Westlaw database contains information from 6 jury verdict reporters, and the LEXIS-NEXIS database contains information from 33 jury verdict reporters.

We searched the Westlaw and LEXIS-NEXIS databases for every jury verdict involving personal injuries due to pathogen-contaminated food between 1988 and 1997. Relevant jury verdicts were identified by searching for the term "food poisoning" or the names of common foodborne pathogens or illnesses (botulism, *Campylobacter,* campylobacteriosis, ciguatera, ciguatoxin, *Clostridium, Cryptosporidium, Cyclospora, E. coli,* hepatitis, *Listeria,* listeriosis, *Salmonella, Shigella, Staphyloccocus, Vibrio, Yersinia, Toxoplasma,* toxoplasmosis, *Trichinella,* and trichinosis). Duplicate cases and cases involving injuries due to causes other than a foodborne pathogen, such as a foreign object in food, were excluded. In some instances, the case summary did not provide enough information to determine whether the plaintiff's illness was allegedly due to a foodborne pathogen. These cases were included in the analysis only if the illness was attributed to a specific food item and produced physical symptoms such as gastrointestinal distress consistent with food poisoning.

After we identified every food poisoning case in the Westlaw and LEXIS-NEXIS databases, we searched the entire LEXIS-NEXIS Litigation Library database for each case to determine whether the case had been appealed to a higher court. We found two cases that had been unsuccessfully appealed by defendants. Several other cases had been remanded for retrial, but we were unable to determine whether these cases had been retried, so they were dropped from the analysis.

The database searches identified a total of 178 jury verdicts from 37 states involving injuries attributed to foodborne pathogens. We compared the verdicts from the areas with full reporting and sample reporting to determine whether there were any systematic differences that might reflect sampling bias in the areas with sample reporting. The comparison was performed by conducting a likelihood ratio test of multivariate models predicting the outcome of trials in each area. The test revealed no significant difference between the populations of verdicts drawn from each area ($\chi^2 = 13.9$, d.f. = 10, $P > 0.05$), suggesting that the sampling was unbiased. Based on this evidence, the cases from areas with sample reporting appear to provide representative information about food poisoning lawsuits resulting in jury verdicts during 1988 to 1997.

A number of different characteristics of each food poisoning lawsuit were coded in a computer database including date of the incident that caused the injury, type of food and pathogen involved, severity of illness, date of trial verdict, outcome of trial, and amount of damage award. Some cases did not report all of this information. Financial damage awards were updated to 1998 dollars using the annual *Consumer Price Index* to ensure comparability between cases.

Results: U.S. Food Poisoning Cases

Based on the 178 food poisoning lawsuits identified by searching the Westlaw and LEXIS-NEXIS databases, food poisoning litigation is a slow process. The average time elapsed between the date of the incident that resulted in illness and the date of the jury verdict was 3.1 years. One case was not tried for nearly 10 years, although another case was tried in just 5 months. The average time required to bring a food poisoning lawsuit to trial was comparable to the average time elapsed between the filing of a product liability lawsuit and the trial verdict (2.5 years) reported in a study of all product liability cases in 5 states during 1983 to 1985 (U.S. GAO, 1989). The slow pace of food poisoning litigation may impose significant costs on consumer plaintiffs. Nevertheless, delay may be advantageous for consumers who developed long-term chronic complications of foodborne illness and were therefore well advised to wait to determine the full extent of their injuries before filing lawsuits specifying monetary damages (Rosenbaum, 1998).

The ability of consumer plaintiffs to identify the specific pathogen and food item that made them ill is likely to have an important effect on the outcome of a trial because of the emphasis placed on establishing a causal link between a defective product and the alleged injury under product liability law. However, less than half (48%) of the food poisoning lawsuits involved a specific foodborne pathogen or illness (Table 6.1). The case summaries for some lawsuits may have failed to record pathogen names, so the actual proportion of lawsuits that involved a specific pathogen

Table 6.1 Foodborne Pathogens Involved in Food Poisoning Lawsuits Decided by Jury Verdicts, 1988 to 1997

Pathogen	*Number of Cases*	*Percent of Cases*[a]
Salmonella (any serotype)	39	21.9
Hepatitis (any type)	10	5.6
Staphylococcus	6	3.4
Vibrio vulnificus	6	3.4
Shigella (any type)	5	2.8
Campylobacter	4	2.2
Mold	4	2.2
E. coli[b]	3	1.7
Botulism (*Clostridium botulinum*)	2	1.1
Ciguatera	2	1.1
Salmonella and *Staphylococcus* combined	1	0.6
Streptococcus	1	0.6
Trichinella spiralis	1	0.6
Vibrio parahaemolyticus	1	0.6
Adverse reaction to protective immunization after exposure to foodborne hepatitis	1	0.6
Not specified	92	51.7
Total	178	100.0

[a]Percents may not add to 100 due to rounding.
[b]The case summaries for the three lawsuits involving *E. coli* did not mention the serotype, but all three cases appeared to involve *E. coli* O157:H7.

might be somewhat higher. Among the lawsuits that named a pathogen, *Salmonella* was the most frequently cited pathogen, followed by hepatitis (any type).

Most lawsuits (92%) identified some kind of food as the cause of illness (Table 6.2). However, one-fourth (26%) of the lawsuits simply named meals such as "dinner" or food categories such as "fast food" that presumably included multiple food items, leaving the precise source of illness unclear. In contrast, two-thirds of the lawsuits (66%) identified a specific food item or type of food as the cause of illness. The most frequently mentioned foods were sandwiches, followed by seafood (excluding oysters) and chicken. Only three lawsuits mentioned packaged items such as canned foods or frozen meals, suggesting that litigation involving packaged foods was either uncommon or else likely to be resolved before trial.

Severity of injuries is a major factor affecting the amount of compensation demanded by consumer plaintiffs. Only five of the food poisoning lawsuits (3%) involved a death, and were brought by family members of the deceased person. All five deaths were elderly persons. Juries are

Table 6.2 Food Items Involved in Food Poisoning Lawsuits Decided by Jury Verdicts, 1988 to 1997

Food Item	Number of Cases	Percent of Cases[a]
Single Vehicle		
Sandwiches (excluding hamburgers and egg sandwiches)	15	8.4
Seafood (excluding oysters)	11	6.2
Chicken	10	5.6
Hamburgers and ground beef	9	5.1
Oysters	9	5.1
Salad	7	3.9
Sausages and unspecified meat	5	2.8
Beverages (excluding milk)	5	2.8
Mexican food	5	2.8
Baked goods (excluding desserts with raw egg)	4	2.2
Chinese food	4	2.2
Packaged meals (*e.g.*, canned food, TV dinner)	3	1.7
Pork	3	1.7
Ice cream	2	1.1
Beef (excluding hamburgers and ground beef)	2	1.1
All other single vehicle (*e.g.*, honey, lasagna)	19	10.8
Multiple vehicle (*e.g.*, restaurant food, fast food, dinner)	46	25.8
Not specified	15	8.4
Total	178	100.0

[a]Percents may not add to 100 due to rounding.

known to be unfavorably disposed towards defendants in cases of wrongful death involving young children. Defendants in such cases may prefer to settle prior to trial. Another one-third of the lawsuits (34%) involved injuries severe enough to require hospitalization. The average length of hospitalization was 9 days, although one plaintiff spent 49 days in the hospital. The remaining lawsuits dealt with less severe injuries that did not require hospital care.

Consumer plaintiffs and defendant firms both used expert witnesses to substantiate their cases. Expert witnesses such as physicians usually receive large consulting fees, and are likely to be called only when their testimony is considered essential. Plaintiffs called one or more physicians as expert witnesses in two-thirds (67%) of the food poisoning lawsuits. In contrast, defendants called physicians in less than half (45%) of the lawsuits. The disparity in the use of medical experts suggests that establishing the role of a foodborne pathogen in causing an illness was a more important issue for plaintiffs than defendants.

Despite their greater reliance on medical experts, most consumer plaintiffs failed to convince juries that defendant firms were legally responsible for causing their illnesses. Only one-third of the food poisoning lawsuits (31%) resulted in a monetary award for the consumer. Unfortunately, only a few of the jury verdict summaries provided commentaries describing why juries decided in favor of firms. Some of the specific reasons cited included failures by consumers to prove that food products were defective, and failure to prove that the consumer actually consumed the food product.

The average amount of compensation awarded to consumer plaintiffs for injuries due to a pathogen-contaminated food product was $133,280 (Table 6.3). However, the distribution of awards was highly skewed because some awards were much larger, including two awards for over $1 million. The median award of $25,560 consequently provides a better indication of the typical award for damages resulting from foodborne illness. The total amount awarded to the 55 consumers who prevailed in court was $7.3 million, but the two largest awards accounted for over half (51%) of this sum.

An alternative measure of the amount received by consumer plaintiffs in food poisoning lawsuits is the expected award, defined by Moller (1996) as the average award multiplied by the probability that the plaintiff will win an award (i.e., 31.4%). The expected award was $41,888, nearly two-thirds larger than the median award. Consumers and firms involved in food poisoning lawsuits can take this expected award into account when making decisions about whether to resolve a lawsuit prior to trial. Consumers could expect to receive this amount if they went to trial. Depending on the court outcome, consumers may also be responsible for their legal and court fees, which typically total about one-third of the award. Conversely, firms could expect to pay this amount in addition to their legal fees and any other costs associated with a public trial, such as loss of business reputation.

Table 6.3 Compensation for Consumer Plaintiffs in Food Poisoning Lawsuits Decided by Jury Verdicts, 1988 to 1997[a]

Average Award	*Median Award*	*Expected Award*
$133,280	$25,560	$41,888

[a]All awards updated to 1998 dollars based on annual *Consumer Price Index*. Calculations exclude 3 cases where the case summary did not report the amount of the award.

Lessons Learned and Future Research Needs

Only one-third of the consumers who filed lawsuits to obtain compensation for injuries due to pathogen-contaminated food products were awarded damages by juries. However, some of these consumers received substantial sums, significantly elevating the expected award above the median award. Food poisoning litigation was also a slow process, and the average lawsuit was not tried until more than three years after the incident that resulted in illness. The lawsuits that were tried involved a wide variety of pathogens and food products, but often did not name a specific pathogen. As a result, many consumer plaintiffs apparently entered court without strong evidence of a causal link between their illness and the defendant's food product. Although a majority of the lawsuits did not involve severe injuries, the remaining lawsuits were conducted on behalf of individuals who were hospitalized or died due to their illness.

Food poisoning lawsuits that went to trial undoubtedly represent only a small and select share of all food poisoning litigation. Consumers and firms both have considerable incentives to settle such lawsuits prior to trial. The incentives for consumers include limiting legal and psychological costs, and avoiding a long delay to an uncertain resolution of the case, which might not be very strong in view of the difficulties involved in identifying the cause of a foodborne illness. The incentives for firms include avoiding the adverse publicity that might result from a public trial, limiting legal costs, and obtaining quick and sure resolution of a potentially costly business problem. The strength of these incentives suggests that lawsuits that went to trial had different characteristics than lawsuits that were settled prior to trial. For example, consumers who went to trial may have overestimated the monetary value of the damages associated with their injuries, while firms that went to trial may have wanted to publicly defend their good names and reputations.

Although reliable estimates of the annual volume of litigation involving foodborne illness are unavailable, it is clear that the vast majority of the 76 million annual foodborne illnesses in the U.S. do not result in lawsuits. Future research should focus on developing a better understanding of the litigation process, because food poisoning lawsuits are a potentially important economic signal to firms to invest more in food safety. Specific questions for research include determining how often lawsuits are filed, how often lawsuits are settled or otherwise resolved before trial, and how settlements differ from court decisions. Consumer complaints and out-of-court settlements are undoubtedly far more frequent than lawsuits that go to trial, and may be the most common signals about the costs to firms for unsafe food.

It is unclear whether foodborne illness litigation will become more frequent in the future. Foodborne illness — and the reasons for litigation — may decrease if firms continue to improve quality control practices to ensure safer food. In contrast, improvements in pathogen detection and identification techniques (including DNA fingerprinting and more rapid microbial tests) may increase the chances that foodborne illness outbreaks will be detected and linked to specific food products and firms. Attorneys who specialize in personal injury cases may also become more interested in handling food poisoning litigation as scientific and technological advances make it easier to link foodborne illnesses to individual firms. In turn, these trends may encourage food firms to further improve quality control standards to reduce the risk of producing contaminated food products that might cause illness and result in litigation.

References

American Law of Products Liability, Third Edition, The Lawyers Co-Operative Publishing Co., Rochester, N.Y., 1987.

Archer, D. L. and Kvenberg, J. E., Incidence and cost of foodborne diarrheal disease in the United States, *J. Food Prot.,* 48, 887, 1985.

Buzby, J. C. and Crutchfield, S. R., New juice regulations underway, *FoodReview,* 22, 2, 1999, pgs. 23–25. <http://www.econ.ag.gov/epubs/pdf/foodrevw/may99/contents.htm>

CAST, Foodborne pathogens: risks and consequences, Task Force Report No. 122, Council for Agricultural Science and Technology, Washington, D.C., Sept., 1994.

Clark, Bruce T., personal communication, Feb., 2000.

Frenzen, P. D., Riggs, T. L., Buzby, J. C., Breuer, T., Roberts, T., Voetsch, D., Reddy, S., and the FoodNet Working Group, *Salmonella* cost estimate updated using FoodNet data, *FoodReview,* 22, 1, 1999, pgs. 10–15. <http://www.econ.ag.gov/epubs/pdf/foodrevw/may99/contents.htm>

Garber, S., Good deterrence, bad deterrence, and challenges in product liability reform, Rand Institute for Civil Justice, 1998. <http://www.rand.org/centers/icj/garber.html>, accessed July 31, 1998.

Gates v. Standard Brands, Inc., 43 Wash. App. 520, 719 P.2d 130 (1986).

Harl, N. E., *Agriculture Law* (through release No. 50), Matthew Bender, New York, Oct. 1997.

Keeton, W. P., Dobbs, D. B., Keeton, R. E., and Owen, D. G. (Eds.), *Prosser and Keaton on The Law of Torts, Fifth Edition,* West Publishing, St. Paul, Minnesota, 1984.

Mead, P. S., Slutsker, L., Dietz, V., McCaig, L. F., Bresee, J. S., Shapiro, C., Griffin, P. M., and Tauxe, R. V., Food-related illness and death in the United States, *Emerging Infectious Diseases,* 5, 5, Sept.-Oct. 1999, pgs. 607–625.

Moller, E. 1996, Trends in Civil Jury Verdicts Since 1985, (Report No. MR-694-ICJ), RAND, Inc., Santa Monica, California, 1996.

Munarriz, R. A., Odwalla, Inc., Daily Trouble, Nov. 25, 1997. <http://www.fool.com/DTrouble/1997/DTrouble971125.htm>, accessed Oct. 14, 1998.

Rasco, B. A., How HACCP can help in a products liability action, *Food Quality,* Oct. 1997, pgs. 16–18.
Restatement of the Law, Second, Torts (Revised & Enlarged), American Law Institute Publishers, St. Paul, Minnesota, 1965.
Roach, J., Odwalla makes a comeback, *Environmental News Network,* Sept. 21, 1999. <http://www.enn.com/enn-features-archive/1999/09/092199/odwalla_5157.asp>, accessed Nov. 1, 1999.
Rosenbaum, D., Food Safety Partners, Northbrook, Illinois, personal communication, July 1998.
The Guide to American Law: Everyone's Legal Encyclopedia, West Publishing Co., St. Paul, Minnesota, 1984.
U.S. Consumer Product Safety Commission, 1997 Annual Report to Congress, U.S. Consumer Products Safety Commission, Washington, D.C., 1998.
U.S. General Accounting Office, Product Liability: Verdicts and Case Resolution in Five States, (Report No. HRD-89–99), Washington, D.C., 1989.

Chapter 7

Consumer Acceptance of Irradiated Meats

John Fox, Christine Bruhn, and Stephen Sapp

Evidence is reviewed from surveys and market studies about the degree of consumer acceptance of food irradiation and it is found that of the studies that provided information about irradiation, almost all provide only favorable information. Results are then presented from two studies that focus on the role of negative information provided by anti-irradiation advocates—one that shows the potentially dominant effect of negative information, and another that suggests that the anti-irradiation message can be effectively countered. Finally, it is discussed how high-involvement consumer decisions about issues such as purchasing irradiated foods are influenced by interactions with family and other social groups and by the viewpoints of opinion leaders. The decision to purchase irradiated food is a socially-constructed decision, and it is concluded that accurate assessment of public reaction to controversial food technologies requires an interdisciplinary effort.

Introduction

Food irradiation is widely used outside the U.S. and has been approved by the U.S. Food and Drug Administration (USDA) for a number of foods, most recently red meats. Despite regulatory approval and virtually unanimous agreement among scientists and their representative organizations

0-8493-2217-0/01/$0.00+$.50

that food irradiation is safe and beneficial, the process has not yet been commercially successful. Food irradiation remains a hard sell to the American public, seemingly in part due to an innate fear of radiation, but also perhaps as a result of information disseminated by opponents of irradiation.

In this chapter, the work on consumer acceptance of irradiation is reviewed with a focus on information provided to subjects. Results are presented from two studies designed to investigate the effect of contradictory positive and negative descriptions of irradiation on consumers' willingness to purchase irradiated meat. In the final section, the broader sociological context within which consumers make "high involvement" purchase decisions is discussed, and it is concluded that while information provided to consumers is certainly important, its effects are tempered or perhaps outweighed by more subtle effects at the family or peer group interaction level.

Consumer Attitudes Toward Irradiated Food

Over the past two decades, several studies have examined U.S. consumer attitudes and willingness to purchase irradiated foods. Because these studies use different information about irradiation, it is difficult to: a) confidently predict market success on the basis of any one study, and b) accurately detect trends toward greater consumer acceptance of the process. For these reasons, consumer attitude surveys are more accurately interpreted by comparing change over time for similar samples given similar information, contrasting attitudes within the same sample, and recognizing that the survey process will cause some concern. For example, when nutrition or food safety concerns are specifically identified, the number of persons expressing concern is two to three times higher than when no topics are identified (opinion research, 1995).

Comparisons within samples suggest that consumer concern about irradiation is lower than other food-related concerns. In a recent nationwide consumer survey in the U.S., 33% of respondents, when specifically asked, identified irradiation as a potential serious health hazard, in contrast to 82% who identified bacteria as a serious hazard, and 66% who classified pesticides as serious (Abt Associates, 1997). In the same study, when consumers were given the opportunity to volunteer food safety concerns, microbiological hazards and spoilage were mentioned by 69% of respondents, but less than one percent mentioned food irradiation. A Gallup study that ranked consumer concern about food processing methods found that irradiation, use of food preservatives, and use of chlorinated water generated similar concern ratings (Gallup et al., 1993). Over time, the number of consumers concerned about irradiation has decreased. In the late 1980s, 42 to 43% of consumers classified irradiation as a serious concern. Those expressing concern decreased in 1992 to 35%, and in 1997 to 33% (opinion research 1987 to 1995, Abt Associates, 1997).

Effect of Educational Information

Consumer studies consistently demonstrate that when provided with science-based information, a high percentage of consumers prefer and are willing to buy irradiated foods (Bruhn et al., 1986b; Bruhn and Schutz, 1989; Gallup et al., 1993). In a simulated market study conducted in Georgia, 44% initially indicated that they would purchase irradiated ground beef. After receiving information on the process, 71% actually selected beef labeled irradiated (Gallup et al., 1993).

The effect of information and product samples on consumer attitudes was documented in a Purdue University study (Pohlman et al., 1994). About half of the sample of 178 residents was willing to buy irradiated foods based upon previous exposure to information about the process. After viewing an eight-minute videotape, *The Future of Food Preservation, Food Irradiation,* subjects demonstrated a significant positive change in knowledge, and willingness to buy irradiated food increased to 90%. Among those who saw the videotape and sampled irradiated strawberries, willingness to buy increased to 99%. These results cannot be generalized to the entire population, since a university community may have a disproportional number of people with more formal education; nevertheless, this study demonstrates high acceptance among specific segments of the population.

The military is considering irradiation to improve the quality of dining hall and field rations. *The Future of Food Preservation, Food Irradiation* videotape and other educational pieces were shown to military personnel (Schutz, 1994). In followup interviews, the percentage of soldiers in the control group who expressed major concern (received no educational intervention) decreased from 33 to 29%, and those expressing no concern increased from 8 to 27%, perhaps due to repeated exposure to the concept of irradiation. Among soldiers who viewed the videotape, 17% expressed major concern and 38% no concern about irradiated food. Those soldiers likely to select irradiated food in the military dinning commons increased from 21% initially to 61% after viewing the videotape. Over 80% indicated they were likely to choose irradiated field rations.

A USDA-funded project in California and Indiana evaluated the impact of a brief educational program on community leaders' attitudes about and knowledge of food irradiation (Bruhn and Mason, 1996). After a brief introduction, the videotape, *The Future of Food Preservation, Food Irradiation* was shown, followed by a question-and-answer period and summary of the effect of irradiation on food. Consumers gained knowledge of specific food irradiation facts, and their interest in purchasing irradiated foods increased. Initially, program participants had little (37%) or no (31%) knowledge about food irradiation, with only 2% believing they had a lot of knowledge. Following the presentation, those believing they were somewhat or very knowledgeable increased to 21 and 59%, respectively.

Characteristics of Acceptors and Rejecters

Innovators lead adoption of new technologies. Early users of new technologies often have higher incomes and lifestyle levels, more prestigious occupations, and more positive self-identities (Rogers, 1971). Although the majority of people respond positively to information about food irradiation, a minority oppose the technology. Those opposed to food irradiation are highly concerned about the use of chemicals on food, place a high value on an "ecologically balanced world," oppose the use of nuclear technology, and prefer to eat only unprocessed or "organic" food (Bruhn et al., 1986a; Bruhn et al., 1987; Brand Group, 1986). Irradiation rejecters have been estimated to be 5 to 10% of the population (Brand Group, 1986).

Demographic factors have been related to views towards irradiation. Women are more concerned about all issues that may affect food safety, including irradiation (Abt Associates, 1996; opinion research, 1987–95; Center for Produce Quality, 1992; Terry and Tabor, 1988; Terry and Tabor, 1990). People with formal education at high school level and above are more likely to purchase irradiated foods (Terry and Tabor, 1990; Resurreccion et al., 1993).

Product Benefits

Applications for Extended Shelf Life, Quality, or Variety

Consumers responded positively to the benefits of irradiation applied to specific products. People were interested in purchasing irradiated tropical fruit, 54%, and irradiated soft fruits, 43% (Schutz et al., 1989). The Food Marketing Institute nationwide survey found 58% of consumers were very or somewhat likely to buy irradiated products to keep them fresh longer (Abt Associates, 1996).

Applications to Enhance Food Safety

About 60% or more consumers prefer irradiated to non-irradiated red meat, pork, poultry, and spices (Schutz et al., 1989; Resurreccion et al., 1995). A nationwide survey conducted by Gallup found that 50% or more of consumers were very or somewhat likely to buy irradiated meat, poultry, and other meat products (Gallup et al., 1993). Additionally, 60% indicated they would pay 10 cents more for irradiated hamburger.

Two economic studies investigated consumer willingness to pay a premium for irradiated products. An Iowa study used an auction technique to investigate consumer reaction to the benefits of irradiating pork (Fox et al., 1993). After auctioning a variety of products, students were given sandwiches made with irradiated or non-irradiated pork, and were given the opportunity to bid up for the product they did not have. The study indicated a very high level of acceptability for irradiated pork in a sample of 58 undergraduate students. Of 29 subjects, 26 paid a premium for irradiated pork to reduce the risk of contracting *Trichina*. Only 1 of 29 students paid to avoid the irradiated product based on an aversion to the irradiation process. Using a similar technique, study participants in Arkansas were willing to pay a premium of $0.75 for a sandwich made with irradiated chicken (Bailey, 1996).

The extensive media coverage of irradiation following the Food and Drug Administration (FDA) approval of the process may have been responsible for increased consumer acceptance of irradiation used to destroy pathogenic microorganisms. In a nationwide study in March 1998, almost 80% of consumers surveyed indicated that they would buy products labeled "irradiated to destroy harmful bacteria" (American Meat Institute, 1998). Sixty-seven percent of consumers said it was "appropriate" to irradiate poultry products, and over 60% felt that the use of irradiated products was appropriate at a fast food restaurant.

Market Experiences

Consumer responses to labeled irradiated food have been positive. In early test markets, irradiated mangoes and papayas sold well (Bruhn and Noel, 1987). A record amount of irradiated strawberries was sold in a Florida produce market in the winter of 1992 (Marcotte, 1992). Since 1992, a small produce and grocery store in the Chicago area has featured irradiated strawberries, grapefruit, juice oranges, and other products (Pszczola, 1992). Owner James Corrigan indicates that irradiated produce outsells non-irradiated by twenty to one or more (Corrigan, 1995).

In 1995, tropical fruit from Hawaii was sold in the Chicago area and several other midwestern markets in collaboration with a study to assess quarantine treatments. Since that time, about 200,000 pounds per year of irradiated fruits from Hawaii including papaya, atemoya, rambutan, lychee, and starfruit have sold in the Midwest and California (Dietz, 1999).

Irradiated poultry has been sold in small markets in Florida, the Chicago area, and Kansas. In a market trial in Kansas in 1995, labeled irradiated poultry captured 60% of the market when priced 10% less than store brand, 39% when priced equally, and 30% when priced

10% higher (Fox and Olson, 1995). The irradiated product sold better in a more upscale store, capturing 73% of the market when priced 10% lower, 58% when priced equally, and 31 and 30% when priced 10 or 20% higher, respectively. This is consistent with other attitude surveys and marketplace data that indicate irradiation is more accepted in affluent markets.

Summary

Attitude and market studies in the U.S. indicate that consumers respond positively to the benefits of irradiation. Consumer questions about irradiation focus on the effect of irradiation on food flavor and wholesomeness, and on worker and environmental safety. Concerns are alleviated through the endorsement of recognized health experts.

The Effects of Positive and Negative Information about Irradiation

As reviewed above, consumer studies over the past 10 to 15 years have indicated a relatively high (usually above 50%) and increasing level of consumer acceptance for food irradiation. However, despite the numerous studies on consumer opinions, sound scientific information on consumer acceptance of food irradiation remains illusive. Because of time and space constraints, telephone and mail surveys typically provide little information to respondents about food irradiation. A limitation of interpreting findings from surveys, therefore, is the difficulty of eliciting thoughtful opinions of food irradiation. A second critical limitation of many studies is that they present only favorable information about food irradiation. Eliciting opinions from a one-sided presentation, regardless of length, can skew research findings. Consider, for example, a study conducted by Market Facts, Inc. on behalf of the Grocery Manufacturers of America. Respondents were presented with this statement about food irradiation:

> Irradiation is a process to sterilize food. It has been approved by the U.S. Food and Drug Administration as a way to make food safer from certain bacteria and slow down spoilage . . . [the statement concludes with a lay-person's explanation of irradiation.]

The study found that 79.5% of respondents said they would likely purchase food labeled "Irradiated to Kill Harmful Bacteria." Now suppose that Food and Water, Inc. had sponsored the survey rather than the Grocery Manufacturers of America. Consider this truthful statement about food irradiation:

> Irradiation is a process that creates cancer-causing chemicals like formaldehyde and benzine in food. Dr. George L. Tritsch of the Roswell Cancer Institute has said, "I am opposed to food irradiation because it is clear that this process increases the levels of mutagens and carcinogens in the food."

Is it reasonable to conclude that this statement would elicit an 80% approval rating of food irradiation? If not, then it should be recognized that conclusions drawn from studies that present only favorable information about food irradiation have limited applicability. As we have said, findings from such studies show responses to favorable scientific information. But because it is unlikely that consumers will hear only favorable information about food irradiation (some argue that university-based researchers have, in fact, an ethical obligation to inform consumers of objections to food irradiation, even though they might not agree with the substance of the objections), it is unreasonable to assume that valid findings on consumer acceptance come from studies that do not present the opinions of both proponents and opponents.

While one of the largest consumer advocacy organizations, the Center for Science in the Public Interest (CSPI), appears to have modified its anti-irradiation stance[1], smaller groups such as Food and Water, Inc., based in Vermont, and the Organic Consumers Association (OCA, formerly the Pure Food Campaign) based in Minnesota remain firm in their opposition. In 1998, these groups almost succeeded in convincing voters in Hawaii to support a ban on food irradiation. The measure failed by only a 1% margin. Consider the following statements from the OCA website in response to the announcement that Titan Corp., with backing from IBP, Excel, Tyson, and other meat processors, would open an electron-beam irradiation facility in Sioux City, Iowa.

> "This plan to use electron beams rather than radioactive material seems to be a public-relations ploy to neutralize opposition to food irradiation. *The food is affected in the same way (vitamins destroyed, new chemicals created, etc.) regardless of the source of the radiation. . . .* Second, the higher-speed electron-beam irradiation can make food slightly radioactive!" (Emphasis in the original.)

[1]In response to the FDA's approval of irradiation for red meat products, CSPI on Dec. 2, 1997, released a statement by executive director Michael Jacobson containing the following: "Irradiation is no silver bullet for improving the safety of meat products. It is a high-tech end-of-the-line solution to contamination problems that can and should be addressed earlier." In Senate testimony dated August 4, 1999 CSPI Director of Food Safety, Caroline Smith DeWaal, in criticizing the structure of the federal food safety system, argued that multiple agencies may prolong the time it takes to bring the benefits of new technologies, among them irradiation for poultry and red meat, to the consumer.

Thus, when irradiated foods from major processors like Tyson do reach the market, they will likely be met by protests and boycotts as part of a campaign to discourage consumers. Given the regionalized nature and relatively limited resources of the opposition, the campaign may be targeted at a small number of urban centers but will nevertheless receive national media coverage. What will be the effect on consumer demand of such a campaign?

Consumer Reaction to Negative Information About Irradiation

Fox et al. (1999) used an experimental market to investigate the way consumers respond to contradictory descriptions of the irradiation process. The experimental design followed that originally used by Shogren et al. (1994), wherein participants are endowed with one good (typically pork) and asked to bid in a repeated, sealed bid, second price auction for an upgrade to an alternative good (irradiated pork). Because experiments use real products and participants actually pay for the good being valued (i.e., for the upgrade from typical to irradiated meat), proponents believe it can produce better estimates of value and more reliable revelation of preferences than a hypothetical mail or telephone survey.

In Fox et al., participants bid for the upgrade to irradiated pork in ten repeated trials. For the first five trials, all participants were provided with the following description of the irradiation process:

> The U.S. Food and Drug Administration (FDA) has recently approved the use of ionizing radiation to control *Trichinella* in pork products and *Salmonella* in poultry. Based on its evaluation of several toxicity studies, the FDA concluded that irradiation of food products at approved levels did not present a toxicological hazard to consumers nor did it adversely effect the nutritional value of the product.
>
> Irradiation of pork products at approved levels results in a 10,000 fold reduction in the viability of *Trichinella* organisms present in the meat.
>
> The forms of ionizing energy used in food processing include gamma rays, x-rays, and accelerated electrons. Ionizing energy works by breaking chemical bonds in organic molecules. When a sufficient number of critical bonds are split in the bacteria and other pests in food, the organisms are killed. The energy levels of the gamma rays, accelerated electrons, and x-rays legally per-

> mitted for processing food do not induce measurable radioactivity in food.
>
> This description is based on a review of the scientific literature on food irradiation.

Given this common informational baseline, 53 participants bid an average of $0.21 for the upgrade to irradiated pork (Figure 7.1). Following the fifth bidding round, participants were allocated to one of three informational treatments—positive, negative, and positive + negative. In the positive treatment, 18 subjects received a favorable description of irradiation based on information provided by the American Council on Science and Health. This description emphasized the benefits and safety of the process and its contribution to controlling foodborne illness. In the negative treatment, 19 participants received an unfavorable description based on information from Food and Water, Inc. This description mentioned that irradiation produced carcinogens called radiolytic products in food, that it lowered the levels of essential vitamins, that it would eliminate warning signs of botulin toxin, and that the use of radioactive materials would put workers and nearby communities at risk. Sixteen participants in the positive + negative treatment were provided with the favorable and unfavorable descriptions simultaneously. Participants in all treatments then proceeded to submit their bids in rounds six through ten without any additional information provided.

Figure 7.1 shows how the different descriptions affected the average bid to obtain irradiated pork. As expected, the favorable description caused bids to increase, and the unfavorable description caused bids to decrease. When subjects were provided with both sets of information,

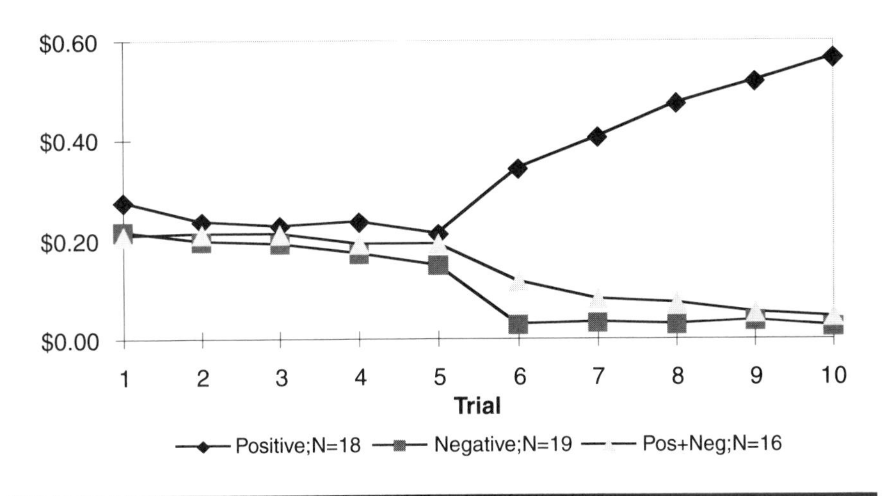

Figure 7.1 Effect of Information on the Average Bid for Irradiated Pork

however, it was clear, given the reduction in the average bid, that the effect of negative information dominated that of the positive. To test the robustness of that finding, Fox et al. repeated the positive + negative experiment with an additional thirty-four subjects and found a similar result—a significant reduction in the average bid. Overall, for 50 subjects, only one submitted a higher bid for irradiated pork after reading both the favorable and unfavorable descriptions.

When subjects were asked which piece of information was most important in causing them to reduce their bid, the most damaging factor was an alleged link between consumption of irradiated food and increased cancer risk. Thus, even if processors use only electron-beam irradiation technology, an anti-irradiation message that focuses only on private risks to consumers to the exclusion of public risks to the environment and workers, retains the potential to be very damaging to consumer confidence in irradiated foods.

Can Negative Information be Countered?

The results described above suggest that an anti-irradiation campaign will negatively affect consumer demand for irradiated products. However, industry stakeholders are unlikely to adopt a stance of merely promoting the benefits of irradiation—they are also likely to counter the typically misleading claims made by opponents. The question, then, is whether the anti-irradiation message can be effectively countered, i.e., whether messages designed to discredit the opponents' claims can reassure consumers, or whether consumers, once they have been exposed to the unfavorable information, will remain unwilling to purchase irradiated products. Here we present results from a study designed to address that question.

In a series of experiments,[2] 96 consumers were asked to make a series of purchase decisions between irradiated and non-irradiated chicken breasts at alternative price levels. For all choices, the price of non-irradiated chicken was held constant at $2.88/lb, while the price of irradiated chicken varied between

- 10% discount to non-irradiated, $2.59/lb
- equal price, $2.88/lb
- 10% premium, $3.17/lb
- 20% premium, $3.45/lb.

[2]These results are based on an extension of experiments reported in Shogren et al., 1999.

Compared to the experiment where participants bid for an irradiated product, in this situation they faced a more familiar task, i.e., choosing which product to buy at the publicly posted prices.

Participants were selected using a random sample purchased from a market research firm. Potential subjects were asked to participate in a consumer economics experiment in return for a participation fee of $25. In order to avoid selection bias, recruiters provided no indication that the experiment was related to food irradiation. On arriving, participants were paid $30 and informed that the experiment would require them to purchase an item costing approximately $5. Following collection of sociodemographic data, the monitor asked participants to read a USDA informational leaflet about food irradiation. Subjects were then shown packages of irradiated and non-irradiated chicken breasts.

Purchase decisions were recorded using a payment card format with four discrete choice questions corresponding to the four prices for irradiated chicken. Subjects were asked to indicate their choice of irradiated or non-irradiated chicken at each price using the following question format.

> You can choose between non-irradiated boneless breasts at **$2.88** per pound, and irradiated boneless breasts also at **$2.88** per pound. Which one **will** you buy?
>
> **I will buy** (Check one)
> ____ **Non-irradiated at $2.88**
> ____ **Irradiated at $2.88**

When all subjects had recorded their decisions, the purchase price for irradiated chicken was determined by randomly drawing one of the four offered prices. Participants then purchased and paid for the product they had chosen at that price. The data revealed that, given the USDA information about irradiation, 80% of subjects chose to purchase irradiated chicken when it was available at the same price as non-irradiated chicken (Figure 7.2). The proportion choosing the irradiated product varied with price in the manner expected, i.e., at the discounted price the proportion choosing irradiated increased to 85%, while at the 10 and 20% premiums the proportion fell to 37 and 14% respectively.

Participants were then provided with additional information about irradiation. In particular, they were provided with the following exact copy of the unfavorable description of irradiation used in the earlier experiment:

> Food irradiation is a process whereby food is exposed to radioactive materials, and receives as much as 300,000 rads of radiation — the equivalent of 30 million chest x-rays — in order to extend the shelf life of the food and kill insects and bacteria.

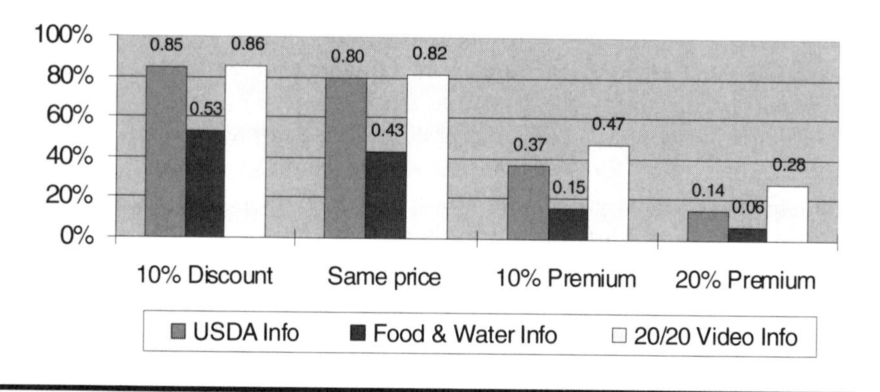

Figure 7.2 Proportion of Consumers Choosing Irradiated Poultry at Alternative Prices

While it is unlikely that food products themselves will become radioactive, irradiation results in the creation of chemicals called radiolytic products in food. Some radiolytic products are known carcinogens. Studies have also suggested that irradiation may be linked to cancer and birth defects. Furthermore, foods exposed to radiation contain lower levels of essential vitamins.

Food irradiation can kill most of the pathogenic bacteria present in food, but so can proper cooking. Moreover, doses of radiation that are adequate to kill *Salmonella* or *Trichinella* are not enough to kill the bacteria that cause botulism. However, such doses would kill the bacteria that signal spoilage through a foul odor. Thus, with irradiation, we would not be able to rely on the usual warning signs that tell us when food is dangerous to eat.

Food irradiation was developed in the 1950's by the Atomic Energy Commission. The objective was to seek potential uses for the byproducts of nuclear weapons production. Today's food irradiation industry is a private, for-profit business enterprise with ties to the U.S. nuclear weapons and nuclear power industries.

Food irradiation also poses potential environmental dangers because of the use of radioactive materials in the process. Workers can be exposed on the job, and entire communities can be exposed in the event of a leak from the plant. Plus, radioactive materials would have to be transported around the country, putting thousands of people at risk in the case of a traffic accident.

This description is based on information supplied by Food and Water, Inc., a consumer advocacy group.

After reading this description, consumers were asked to indicate what their purchase decision *would be* if they were allowed to repeat it. Decisions made under this hypothetical scenario were recorded in the same way as the purchase decisions, i.e., by responding to four discrete choice questions corresponding to the four prices for the irradiated product. Figure 7.2 shows the negative impact of the Food and Water, Inc. information on demand for the irradiated product. At equal prices, the proportion choosing irradiated chicken fell from 80 to 43%, and only 15% were willing to pay a premium. The question we now addressed was whether additional information, favorable to the process and discrediting the claims made by opponents, would restore consumer confidence.

In an attempt to counter the opponents' claims, subjects were shown a report on food irradiation, hosted by John Stossel of ABC News for the 20/20 news program. The report, entitled *The Power of Fear*, shown on December 13, 1991, focused on protests at a newly constructed food irradiation facility in Florida. The host interviewed the plant's developer and representatives of Food and Water, Inc. who were instrumental in organizing and leading the protest. The conclusions reached by the host were: a) that food irradiation was a safe process and that he would prefer irradiated to non-irradiated foods given the choice, and b) that claims made by opponents of the process were, at best, misleading or based on irrelevant science. At the conclusion of the 20/20 video, the monitor discussed each claim about irradiation made in the unfavorable description. The monitor emphasized the following points:

- that irradiated food products would never become radioactive
- that radiolytic products, similar to those produced by irradiation, were also produced when foods were grilled or fried
- that no studies had shown a connection between food irradiation and cancer or birth defects
- that vitamin losses were insignificant and lower than those associated with other food processes such as canning or freezing
- that irradiation at approved doses did not sterilize food and thus the spoilage warning signs were not lost
- that there were no links between food irradiation and nuclear weapons or nuclear power
- that irradiation had been used to sterilize medical devices and several

> consumer products other than foods for several decades with no problems related to the use or transportation of radioactive materials.

Once again, consumers were asked to indicate what their purchase decision *would be* if they were now allowed to repeat it. Figure 7.2 shows that, having viewed the 20/20 video and listened to the pro-irradiation counter claims, consumer confidence in the process appeared to have been fully restored. At equal prices, the proportion now saying they would choose irradiated chicken breasts was 82%, with 47% willing to pay at least a 10% premium. The anti-irradiation message was effectively countered.

Because revealed preferences are responsive to new information, a relevant question is "who reacts how?" Based on the purchase decision data, we can classify consumers into three types: a) proponents (consumers always indicating a preference for the irradiated product when available at an equal price), b) opponents (consumers always preferring the non-irradiated product at equal prices), and c) for want of a better descriptor, "swing voters" (consumers whose choice at equal prices changes with new information). Figure 7.3 describes the changes in preferences as new information was introduced. Of the 76 subjects initially preferring irradiated chicken, 40 continued to prefer irradiated after being exposed to the unfavorable information. These 40 constitute the class of proponents. All 20 subjects who initially chose non-irradiated continued to do so given the unfavorable information, but following the video and discussion, 7 of the 20 switched and indicated a preference for irradiated chicken. The remaining 13 constitute the class of opponents. The "swing voters" number 43.

While any such classification must be arbitrary (different criteria leading to different classifications), this one is complete, has some logical appeal, and allows us to ask (and hopefully provide some answers to) the above question. To analyze the data, we used a multinomial logit model, where a participant's classification into a given category is modeled as a function of his or her characteristics. Our demographic data included

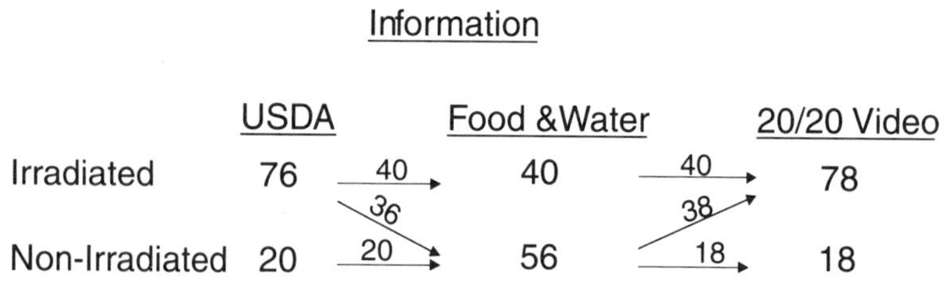

Information

USDA Food &Water 20/20 Video
Irradiated 76 40 40 40 78
36 38
Non-Irradiated 20 20 56 18 18

Figure 7.3 Participants Choosing Irradiated Poultry at Equal Price given New Information

information about gender, age, household members, education level, and income. Additionally, we asked respondents to estimate the risk of illness from consuming non-irradiated chicken by responding to the following question:

> If one million people ate typical chicken, how many do you think would become ill from *Salmonella*? _____

Table 7.1 shows the estimated marginal effects (the effect of a one-unit increase in the independent variable on the probability of a respondent entering each classification). Note that across categories, marginal effects must sum to zero. Given the relatively small sample, few of the estimated effects are statistically significant at the traditionally accepted levels. Nevertheless, some of the results are of interest. First, males are greater than 20% more likely than females to be classified as a proponent of the process. This finding is consistent with that of several other studies that found females more concerned about and less likely to purchase irradiated foods (Sapp et al., 1995; Malone, 1990). Second, more highly educated consumers are more likely to be either opponents or proponents and less likely to be classified as "swing voters." Most studies find a positive relationship between education and acceptance of irradiation. Third, the presence of children in a household is significantly associated with respondents being classified as opponents of irradiation. Fourth, as

Table 7.1 Marginal Effects of Independent Variables on Respondent Classification

	Classification		
Variables	*Opponent*	*"Swing Voter"*	*Proponent*
Constant	−0.192	0.453*	−0.261
	(−1.42)[a]	(1.86)	(−1.08)
Male	−0.063	−0.163	0.226*
	(−0.85)	(−1.36)	(1.92)
Age	−0.001	0.000	0.001
	(−0.54)	(0.00)	(0.44)
Education	0.019	−0.046	0.027
	(1.00)	(−1.31)	(0.76)
Kids in Household	0.110*	−0.089	−0.021
	(1.71)	(−0.70)	(−0.16)
Salmonella Risk	−0.36 E-06	−0.54 E-06	0.90 E-06
	(−0.64)	(−0.89)	(1.57)

[a]Values in parentheses are t-values.
*Statistically significant at the 0.10 level.

expected, the higher the respondent perceives the risk of illness from non-irradiated chicken, the more likely he or she is to be a proponent of irradiation. Finally, as found in other studies, age of the respondent has no consistent effect. Given the small, local sample used for this study, we cannot generalize from these results, but if the findings were to be confirmed in other studies, they suggest that a prime target audience for a pro-irradiation educational campaign would be females with lower levels of education in households with children.

Socially Constructed Risk Assessments

Much research on consumer opinions has focused on how best to educate people about food irradiation. Although thorough consumer education is necessary, it is unlikely to be sufficient input for consumers to adopt irradiated food. The reason for this gap between knowledge acquisition and behavior change is that decisions about complex and controversial technology rely upon a very broad spectrum of inputs in addition to scientific facts. Social scientists describe decisions about complex and controversial technologies as "socially constructed" decisions.

Consider a consumer with a high-involvement decision to make, such as whether to adopt food irradiation. We are defining "high-involvement" from the perspective of the consumer. That is, the consumer might not know the toxicological risk associated with eating irradiated food but knows that making the right call on what food to eat is important, and therefore demanding of one's full attention (an example of a low-involvement decision, in contrast, would be deciding which color toothbrush to buy). Our consumer is unlikely to pursue the same *type* of decision-making process for food irradiation as he/she would do for a low-involvement decision. For the high-involvement decision, the individual will consult with family, friends, and other trusted people, such as physicians. The individual will be highly influenced by the viewpoints of "opinion leaders," trusted people and/or organizations who carry much prestige in the area of the technology (Rogers, 1995). In short, high-involvement decisions entail a much more complicated process than low-involvement decisions and are influenced strongly by interactions with others and the viewpoints of opinion leaders.

Consequently, opinions of complex and controversial technologies, such as food irradiation, are influenced by a wide array of scientific, economic, social, and ethical issues, and shift with changing current events. Gaining adoption of these technologies, therefore, can be a frustrating, time-consuming activity. What role does education about the technology play in adoption of high-risk technology? Although information about technology is an essential element of decision-making, it is not nearly sufficient to bring about adoption of high-risk technology.

As demonstrated by Bord and O'Connor (1989) and Sapp and Harrod (1994a), precisely because it is such a complex and controversial topic, consumer acceptance of food irradiation is inherently a socially constructed decision. That is, persons must gauge the opinions of others regarding food irradiation before they can develop well-formed opinions of it themselves. Sapp and Harrod (1990), for example, found that negative comments made during group discussion influenced opinions more so than favorable comments, which implied that normative factors can be important determinants of consumer acceptance of food irradiation. Sapp and Harrod (1994a) found that the opinions of subjects who were allowed to discuss irradiation in groups following education about it were more extreme than those of subjects who were not allowed discussion time. Sapp and Harrod (1994b) found attitude-behavior consistency for subjects who discussed food irradiation before stating their opinions of and intentions to eat irradiated food, but not for subjects who stated their opinions and intentions without prior discussion. These findings support a socially constructed perspective of risk assessments which emphasizes the importance of many economic, social, political, and cultural factors other than toxicological evidence on opinions of food irradiation.

In fact, studies show that the most important factor affecting adoption of food irradiation is trust in government and industry. Trust accounted for 74% of the explained variance, and accuracy of knowledge accounted for an additional 20% of the explained variance in consumer acceptance of food irradiation in a study conducted by Bord and O'Conner (1989). And Sapp et al. (1995) found that trust in government and industry was the most important variable in predicting responses to four measures of acceptance. Similarly, Sapp et al. (1995) found that word of mouth and trust in government and industry were much more important than demographic factors in predicting consumer acceptance of irradiation.

Thus, given the complex and controversial nature of the debate about food irradiation, surveys and laboratory studies are not adequate mechanisms for understanding or predicting consumer acceptance of food irradiation. The *only* effective way of predicting consumer behavior when irradiated food is on the grocery shelves is to conduct a market test, wherein consumers are allowed to hear both pro- and anti-irradiation messages, discuss food irradiation with friends, family, and trusted experts such as their health-care providers, and then offer their opinions of and intentions to purchase irradiated food.

References

Abt Associates Inc., Food Industry & Agribusiness Consulting Practice, 1996, 1997, *Trends in the United States, Consumer Attitudes and the Supermarket,* Food Marketing Institute, Washington, D.C.

American Meat Institute, 1998, New consumer research good news for irradiated foods, American Meat Institute, Document #16719, Washington, D.C.

Bailey, W.C., 1996. Comparative study of the willingness to pay for organic and irradiated meat products: an experimental design, *Consumer Interests Annual,* 42:1–5.

Bord, Richard J. and O'Conner, Robert E., 1989, Who wants irradiated food? Untangling complex public opinion, *Food Technology,* 43: 87.

Brand Group, 1986, Irradiated seafood product. A position paper for the seafood industry, final report, Chicago, IL.

Bruhn C.M. and Mason, A., 1996, Science and society: a public information program on food innovations, final report USDA FY 1994 Special Projects, Project no 94-EFSQ-1–4141, U.S. Department of Agriculture, Washington, D.C.

Bruhn C.M. and Noel, J.W., 1987, Consumer in-store response to irradiated papayas, *Food Technology,* 41(9): 83.

Bruhn, C.M., and Schutz, H.G., 1989, Consumer awareness and outlook for acceptance of food irradiation, *Food Technology,* 43(7):93–94, 97.

Bruhn, C.M., Schutz, H.G., and Sommer, R., 1986a, Attitude change toward food irradiation among conventional and alternative consumers, *Food Technology,* 40(1):86–91.

Bruhn, C.M., Schutz, H.G., and Sommer, R., 1987, Food irradiation and consumer values, *Ecology of Food and Nutrition,* 21: 219.

Bruhn, C.M., Sommer, R., and Schutz, H.G., 1986b, Effect of an educational pamphlet and posters on attitude toward food irradiation, *Journal of Industrial Irradiation Technology,* 4(1):1.

Center for Produce Quality, 1992, Fading scares—future concerns: trends in consumer attitudes toward food safety, Produce Marketing Association, Alexandria, VA.

Center for Science in the Public Interest, Testimony of Caroline Smith DeWaal before the Senate Committee on Governmental Affairs on *Overlap and Duplication in the Federal Food Safety System,* Washington, D.C., August 4, 1999. www.cspinet.org/foodsafety/ffss.html.

Center for Science in the Public Interest, Statement of Michael Jacobson, executive director, on *FDA's Approval of Irradiation for Red Meat Products,* December 2, 1997. www.cspinet.org/new/mikeirad.htm.

Corrigan, J., 1995, personal communication, Carrot Top Market, Illinois.

Dietz, G., 1999, personal communication, Isomedics.

Fox, J.A., Hayes, D.J., Kliebenstein, J.B., Olson, D.G., and Shogren, J.F., 1993, The acceptability of irradiated meat, poster presented at Valuing Food Safety and Nutrition Workshop, Alexandria, Virginia, June 2–4.

Fox, J.A., Hayes, D.J., and Shogren, J.F., 1999, Consumer Preferences for Food Irradiation: How Favorable and Unfavorable Descriptions Affect Preferences for Irradiated Pork in Experimental Auctions, Kansas State University, working paper.

Fox, J.A. and Olson, D.G., 1998, Market Trials of Irradiated Chicken, *Radiation Physics and Chemistry,* 52:63–66.

Gallup Organization, Abt Associates, Center for Food Safety and Quality Enhancement, University of Georgia, 1993, Consumer awareness, knowledge and acceptance of food irradiation, American Meat Institute Foundation, Arlington, VA.

Malone, J.W. Jr., 1990, Consumer willingness to purchase and to pay more for potential benefits of irradiated fresh food products, *Agribusiness,* 6:163–177.

Marcotte, M., 1992, Irradiated strawberries enter the U.S. market, *Food Technology,* 46(5): 80.

Opinion research, Trends. Consumer attitudes and the super-market, Food Marketing Institute, Washington, D.C., 1987–1995.

Organic Consumers Association, statement September 27, 1999, www.purefood.org/Irrad/hispeedelec.cfm.

Pohlman, A.J., Wood, O.B., and Mason, A.C., 1994, Influence of audiovisuals and food samples on consumer acceptance of food irradiation, *Food Technology,* 48(12):46–49.

Pszczola, D.E., 1992, Irradiated produce reaches midwest market, *Food Technology,* 46(5): 89.

Resurreccion, A.V.A., Galvez, F.C.F., Fletcher, S.M., and Misra, S.K., 1995, Consumer attitudes toward irradiated food: results of a new study, *Journal of Food Protection,* 58(2):193–196.

Rogers, E.M., 1995, Diffusion of Innovations, Fourth Ed., New York, The Free Press.

Rogers, E.M. and Shoemaker, F.F., 1971, *Communication of Innovations,* New York, The Free Press, p. 27.

Sapp, S.G. and Harrod, W.J., 1990, Consumer acceptance of irradiated food: an examination of symbolic adoption, *Journal of Home Economics and Consumer Studies,* 14:133–45.

Sapp, S.G. and Harrod, W.J., 1994, The social construction of consumer risk assessments, *Journal of Home Economics and Consumer Studies,* 18:97–106.

Sapp, S.G. and Harrod, W.J., 1994, Socially constructed subjective norms and Subjective Norm-Behavior Consistency, *Social Behavior and Personality,* 22(1):31–40.

Sapp, S.G., Harrod, W.J., and Zhao, L., 1995, Social demographics and attitudinal determinants of consumer acceptance of food irradiation, *Agribusiness,* 11(2):117–130.

Schutz, H.G., 1994, Consumer/Soldier acceptance of irradiated food, U.S. Army Natick Reseach, Development, and Engineering Center, Contract # DAALO3–91-C-0034, Natick, MA.

Schutz, H.G., Bruhn, C.M., and Diaz-Knauf, K.V., 1989, Consumer attitudes toward irradiated foods: effects of labeling and benefits information, *Food Technology,* 43(10):80–86.

Shogren, J.F., Shin, S.Y., Hayes, D.J., and Kliebenstein, J.B., 1994, Resolving differences in willingness to pay and willingness to accept, *American Economic Review,* 84:255–270.

Shogren, J.F., Fox, J.A., Hayes, D.J., and Roosen, J., 1999, Observed choices for food safety in retail, survey, and auction markets, *American Journal of Agricultural Economics,* 81:1192–1199.

Terry, D.E. and Tabor, R.L., 1988, Consumer acceptance of irradiated produce: a value added approach, *Journal of Food Distribution Research,* February, 73–89.

Terry, D.E. and Tabor, R.L., 1990, Consumers' perceptions and willingness to pay for food irradiation, presented at the Proceedings of the Second International Conference on Research in the Consumer Interest, August 1990.

Chapter 8

Evaluating Impact of Food Safety Control on Retail Butchers

Adrian Peters, Matthew Mortlock, Chris Griffith, and David Lloyd

The Pennington Report into an outbreak of Escherichia coli *O157 food poisoning, linked to the supply of cooked meats by a retail butcher, recommended implementation of the Hazard Analysis Critical Control Point (HACCP) system in all U.K. butchers. Following this, initiatives were launched in England and Wales to assist HACCP implementation in retail butchers. While the increased adoption of HACCP is to be applauded, there has been relatively little research evaluating HACCP implementation or benefits and economic impact arising therefrom. This chapter reports on evaluation of the HACCP initiatives for retailer butchers in Wales and England, U.K.*

Introduction

In spite of years of scientific research, food safety continues to be a problem worldwide. It has been recently stated, "millions of words of advice and exhortation have been said and written on the subject, but the problem seems steadily to be getting worse instead of better" (Wollen, 1999).

0-8493-2217-0/01/$0.00+$.50

One response increasingly adopted has been to establish a government department (food standards agency or equivalent) with specific responsibility for food safety. Such organizations are being talked about throughout Europe and as far afield as Australia and New Zealand, each embracing the "farm to fork" philosophy.

If such concern about food safety exists and research has been ongoing for many years coupled with improvements in legislation, it is difficult to explain why the problem of unsafe food remains. Figure 8.1 illustrates one way in which food safety can be broken down into component subjects and relevant associated disciplines, although it is recognized that there may be much overlap between these areas.

Much, and probably most, of the early and present work on food safety has focused on the vertical axis of Figure 8.1. Scientists have for years used chemistry, physics, microbiology, and related disciplines to study the nature of hazards and how food processing affects them. The dominance of this classical food science approach is reflected by the number of journals reporting the results of published work, and is illustrated in the proceedings of the Food Microbiology and Food Safety into the Next Millennium Conference held in Eindhoven, The Netherlands, in 1999. The proceedings were condensed into an excellent volume (various authors, 1999) of over 900 papers, with sections on preservation, mycology, microbial physiology, etc., but with no mention of people, and little mention of systems.

However, perhaps the dominance of the physical/biological science approach is one of the reasons food safety remains a problem. Concentra-

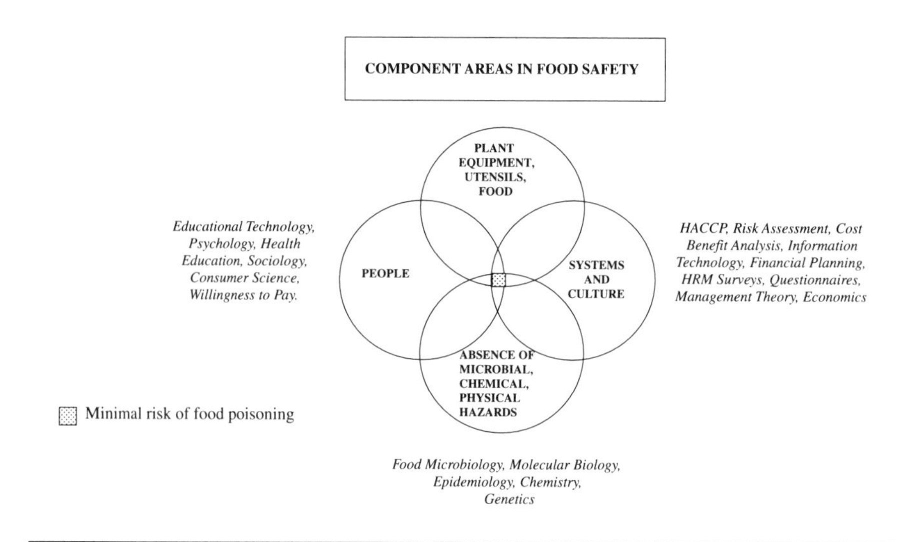

Figure 8.1 Component Areas in Food Safety

tion on research and development along the vertical axis (science) may have been at the expense of subjects on the horizontal axis (management). It has been reported that poor food handling practices contribute to 97% of food illnesses in food service establishments and the home (Howes et al., 1996), with the majority of food-related illnesses likely to be attributed to these two locations (Griffith, 2000). This highlights the importance of people and training in the prevention of food poisoning, although how many of these poor food-handling practices can be attributed directly to human error, as opposed to systems or cultural failure, is not known. Perhaps there needs to be a shift in emphasis from molecular biology to better implementation of the food safety practices that are known. It is certainly true that more and more attention is being given to Hazard Analysis Critical Control Point (HACCP), which is a food safety management system, and this is to be applauded. However, to date there has been relatively little evaluation of the process of implementation and the long term benefits achieved by such systems. Similarly, while hygiene and HACCP training are often recognized as having a vital role, there has been relatively little research assessing or improving their efficacy. It was a perceived need to try to take a fresh look at how food safety is impacted by people and management areas (horizontal axis) as well as the scientific problems (vertical axis), which led to the formation of the Food Research and Consultancy Unit at the University of Wales Institute, Cardiff (UWIC).

Background

The Food Research and Consultancy Unit at UWIC consists of a multidisciplinary food safety research group, a food industry center (FIC) and a biotechnology support group. Figure 8.2 recognizes the importance of interdisciplinary links and their contribution to food safety throughout the food chain.

The Food Safety Research Group includes the study of food safety in manufacturing, food service establishments, and the home. It uses traditional scientific as well as novel techniques (Table 8.1). The FIC is specifically designed to help small and medium-sized enterprises (SME) and recognizes that this type of company frequently does not have access to the latest scientific information on food safety. Communicating and linking research with industry is therefore a key role of the FIC. As well as training, problem solving and linking with industry via specialist government schemes (Teaching Company Schemes—TCS) are essential elements of the work of the FIC. In addition, the Biotechnology Support Unit does not help the food industry directly, but conducts research useful to companies that do. This unit has, for example, in the past year published work on ATP bioluminescence and hygiene monitoring (Davidson et al.,

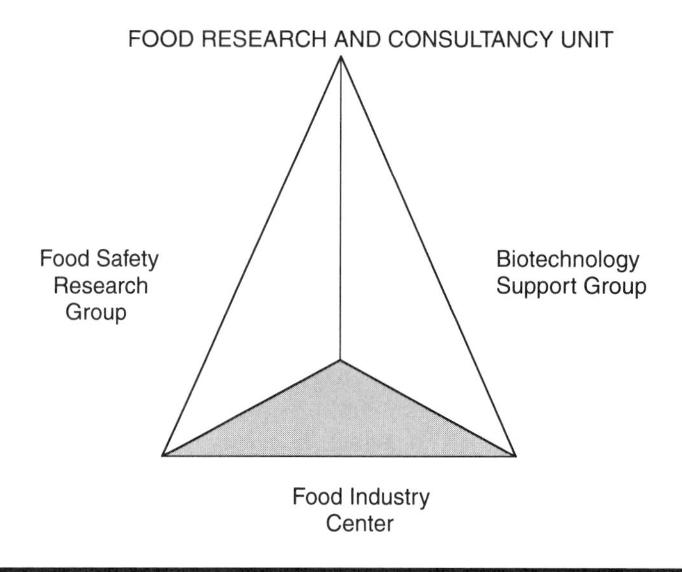

Figure 8.2 Food Research and Consultancy unit

1999), as well as the bactericidal potential of ozone as a terminal disinfectant (Moore, et al., 2000), both in relation to use in the food industry.

A food safety research unit study of barriers to the implementation of food hygiene management systems in the U.K. food industry funded by the U.K. Ministry of Agriculture, Fisheries, and Food (Peters et al., 2000) included a comprehensive survey of hygiene management practices in the manufacturing, food service, and catering sectors. Conducted in mid-1997, the research highlighted the differences in the level of implementation of hygiene management systems across the sectors, with retail food busi-

Table 8.1 Techniques Used by Food Safety Research Group

Technique	*Recent References*
Microbiological analysis	Harrison et al., 2000
Notational analysis	Griffith et al., 1999
Social marketing	Redmond et al., 1999
Social cognition models	Clayton et al., 2000
Human error analysis	
Surveys/Questionnaires	Mortlock et al., 1999
Case studies	
Cost benefit analysis	Mortlock et al., 2000b
Willingness to pay	Peters et al., 2000
Risk assessment/Exposure assessment	Griffith et al., 1998
Quantitative food safety auditing	Coleman and Griffith, 1998

nesses being significantly less likely to have adopted HACCP (Mortlock et al., 1999). Businesses also underestimated the risk they posed to the consumer, and this influenced the likelihood of HACCP implementation. Major barriers to change included the perceived cost and difficulty of obtaining expert advice, and the lack of tangible benefits arising from the use of HACCP, particularly among SME in the food service and retail sectors. Training costs and the lack of suitably trained staff at all levels also influenced HACCP use, and suggested that training provision would assist in extending HACCP use across the food industry (Mortlock et al., 2000a).

In December 1996, an outbreak of enterohemorrhagic *Escherichia coli* food poisoning in Lancashire, Scotland affected over 400 people and resulted in 21 deaths. Linked to the supply of cooked meats by a retail butcher, the outbreak prompted a government report produced by Professor Hugh Pennington of Aberdeen University (Pennington, 1997). The Pennington Report made a number of recommendations, including implementation of all seven principles of HACCP in retail butchers. Following this, initiatives were launched in England and Wales to provide accelerated HACCP implementation in the sector. Both initiatives focused on SME. Major supermarket chains were not included, as they had already implemented HACCP programs.

In January 1998, the Food Research and Consultancy Unit at UWIC approached representatives from the then Welsh Office (now the National Assembly for Wales) and the Welsh Food Safety Technical Panel (made up of representatives from the National Assembly and councils across Wales) to discuss the potential for a joint research project. Aware of the development of the accelerated HACCP initiative in Wales, the unit had identified the specific need to carry out an all-Wales, research-based evaluation of the initiative.

During the development and planning for the Welsh research, the English HACCP initiative was also launched, managed by the Meat and Livestock Commission (MLC), a non-departmental government-funded public agency representing the interests of the U.K. meat industry. In January 1999, the MLC were contacted to discuss the possibility of a parallel study to the one in Wales. This was agreed in February 1999, and a methodology was developed, with several notable differences to the research in Wales.

The main aims of the research were to:

- gather basic background data on the butchery industry and its practices prior to being involved in the HACCP initiative
- estimate the costs to businesses of their involvement in the initiative and implementation of HACCP
- assess the attitudes and beliefs of business managers towards the HACCP system and the factors that motivated them to take part in the initiative

- evaluate the impact of the training provided upon managerial knowledge and ability relating to HACCP and food safety in general.

Methods

Although the research aims were the same for both the Welsh and English initiatives, different methodologies were adopted. The Welsh study was completed before the English study, and this informed the experimental approach. In addition, the larger population of retail butchers in the English study facilitated a modified approach.

Welsh HACCP Initiative

Local environmental health departments in each of the 22 unitary authorities that form the framework for local government across Wales were responsible for the delivery and management of the Welsh Accelerated HACCP initiative. After over a year of negotiation with the Welsh Food Safety Technical Panel and questionnaire piloting, 19 authorities accepted the invitation to take part in the research, yielding a population of 730 businesses for the main study.

Questionnaire Design and Administration

Both the questionnaire design and the methodological approach adopted for the administration of the survey were developed in consultation with the Food Safety Technical Panel, including a pilot study on a small sample of butchers in three South Wales councils. While the pilot suggested that the ideal method would have been for councils to deliver the questionnaires by hand, concerns about the time commitment required meant that this approach was not able to be prescribed across the board. Instead, it was left to the discretion of individual departments as to how to distribute the questionnaires, either by post or in person by local Environmental Health Officers (EHO) in the course of normal duties.

Administration of the questionnaire would have been more straightforward had it been carried out centrally by UWIC, particularly regarding following up non-respondents. However, the Food Safety Technical Panel felt that councils would prefer a more hands-on role and greater ownership of the research. Local administration of the questionnaires also ensured the anonymity of the individual businesses involved in the study. This anonymity was deemed important in helping to improve the quantity and reliability of responses.

The questionnaire was aimed at business or technical managers in retail butcher shops and sought responses on business demographics, business practices with regard to both prerequisite hygiene programs (Good Hygiene Practice), and implementation of specific HACCP principles prior to the training initiative. The questionnaire also asked respondents to retrospectively indicate the cost of implementing and running their HACCP system. Direct costs and time spent on each activity were estimated. Finally, attitude statements were used to assess those factors that affected their adoption of HACCP.

English HACCP Initiative

Two separate questionnaires were designed for the English study, each the result of modifications made to the single questionnaire used in Wales. These focused on separate areas of interest, one on the attitudes of business managers and improvements in their knowledge and ability as a result of HACCP training, the other on the economic impact of HACCP implementation. This approach was made possible by the larger sample available for the English study, with 500 of each questionnaire sent out to the first 1000 businesses to complete the implementation of HACCP via the English Initiative. The only drawback of this approach was that separating the issues of costs and managerial attitudes prevented any analysis of the relationships between these two variables.

Questionnaires were sent out at the beginning of September 1999 to the first 1000 businesses identified to have completed their HACCP implementation, using address labels provided by the MLC. 500 businesses received questionnaires focusing upon managerial attitudes and the impact of training upon knowledge and ability, the other 500 assessing cost and time issues associated with implementing the HACCP system. Completed questionnaires were returned directly to UWIC, using pre-paid envelopes that had been provided. Three weeks after the initial mail shot, all 1000 businesses received a reminder/thank-you letter, and while response rates had already been good up to this point, these had a further positive impact on the response.

Questionnaire Design

Due to the time involved in developing the research, a direct before-and after-comparison of hygiene issues was not feasible. Nevertheless, respondents were asked to retrospectively identify their own business practices and the levels of knowledge and ability they held prior to the Initiative. Attitude scales were used to identify factors affecting HACCP implementation.

Economic Impact Assessment

The costing questionnaire examined the direct costs and time spent on various activities relating to HACCP. These were estimated retrospectively, and covered the same activities as the Welsh study.

Data Analysis

Questionnaire responses were coded and entered into SPSS for Windows for statistical analysis. Chi-square (χ^2) was used to test for the presence and nature of significant associations for nominal or categorized data, while Mann Whitney U tests or independent/paired sample "t" tests were used assess interval data. Significance is reported at the 5% level or lower.

The collation and analysis of costing data required the use of a sophisticated database. Such a database was developed in conjunction with M.D. Associates, Grimsby, U.K. for a Ministry of Agriculture, Fisheries, and Food funded study of the costs and benefits of HACCP adoption (Mortlock et al., 2000b). The database was designed to be sufficiently flexible to accommodate a range of cost data from different industry sectors.

Database Design

The database allowed data entry under three main headings: HACCP planning and implementation costs, ongoing running costs (recurring costs), and post-HACCP costs. Within each section, a number of cost centers was established to mirror the design of the costing questionnaires. Data was input as direct cost or time spent on an activity.

A global data input screen allowed the user to define hours per week for full and part-time staff and a global rate of pay. For this study, the average value to the business of each person-hour spent on the activities listed was estimated at £5.50, using figures provided by the Office of National Statistics from the New Earnings Survey for 1998 (ONS, 1998)[1].

A demographics input screen allowed baseline data for each business entered. This included business size and locality (urban, suburban, rural).

[1]£1.00 Sterling is approximately equal to $1.50 U.S. (June 2000).

Results

Response Rates

In the Welsh study, a total of 135 responses was received—an overall response rate of 18.5%, although 6 unusable responses resulted in a total valid response of 17.7%. The anonymity of responses did not allow a follow-up study of non-respondents, so it is impossible to determine whether response was biased, although the range of attitudes expressed in the questionnaire do not suggest this.

In the English study, a total of 334 unspoiled responses was received, a response rate of 33.4%. This rate varied between the two questionnaires, with the attitudinal questionnaire achieving a response rate of 38.2% and the costing questionnaire 28.6%. It is clear that the greater complexity of the economic assessment questions affected the response, and that the strategy of splitting the questionnaire led to a greater overall response. There was no follow-up study of non-respondents, but again, the range of attitudes expressed does not suggest that there was any undue bias.

Baseline Data

Business size was assessed using the number of full and part-time food handlers employed. A food handler was defined for this study as anyone who handled or prepared food whether open/unwrapped or packaged. Business turnover was also used as a measure of business size. Table 8.2 shows business size for Welsh and English butchers. The results show that the retail butcher sector is dominated by small businesses, and that there is no difference in the baseline characteristics of businesses in Wales or England.

Table 8.2 Business Size in Terms of Food Handlers and Turnover in Welsh and English Retail Butchers

	Welsh Butchers	*English Butchers*
Food handler mean	3.2	3.8
50 percentile	2.5	3
75 percentile	4	5
Food handler max.	14	30
Weekly turnover mean	£5,082	£4,030
50 percentile	£3,000	£3,000
75 percentile	< £5,082	< £5,000
Weekly turnover max.	£40,000	£32,000

Table 8.3 Location of Retail Outlets Welsh and English Retail Butchers

	Welsh Butchers	*English Butchers*
Urban	11%	16%
Suburban	37%	45%
Rural	52%	39%

Table 8.3 shows the geographical location of businesses responding to the survey. There was no significant difference between the Welsh and English businesses, although the slightly higher proportion of businesses in rural areas in Wales reflects the largely rural nature of much of Mid, West, and North Wales.

The business status of survey respondents is shown in Table 8.4. The majority of response was from independent butcher shops. While this may not reflect the industry as a whole, as the major supermarket chains were not included in the study, it nevertheless demonstrates the prevailing nature of business ownership among butchers in Wales and England. The similarity in baseline data for the two studies provides a level of assurance that the slightly different data collection methods employed did not lead to a particular response bias.

Business Practices Prior to the HACCP Initiatives

Table 8.5 shows the hygiene practices (prerequisite programs) and HACCP principles used in businesses before the HACCP initiatives. This data is based on managers' recollections and opinions, and no attempt was made to audit businesses.

While most businesses practiced stock control, inspection of incoming foods, and temperature monitoring of foods, fewer had pest control programs, temperature monitoring of equipment, or cleaning schedules in place prior to the HACCP initiatives. HACCP was not widely used before

Table 8.4 Business Status of Survey Respondents

	Welsh Butchers *n = 129*	*English Butchers* *n = 334*
Independent shop	74%	88%
Market stall	5%	2%
Part of local chain	4%	5%
Butcher/deli counter in larger store	12%	5%
Catering butcher	5%	0%

Table 8.5 Prerequisite Programs and HACCP Principles Employed before the HACCP Initiatives

	Welsh Butchers *n = 129*	*English Butchers* *n = 334*
Cleaning schedules	77%	65%
Pest control programs	57%	61%
Microbiological testing	4%	6%
Temperature monitoring (food)	78%	81%
Temperature monitoring (equipment)	59%	62%
Inspection of food deliveries	88%	87%
Food hygiene training	68%	84%
Stock rotation	96%	95%
Hazard analysis	63%	60%
Identification of CCPs	56%	44%
Target levels/critical limits	43%	40%
Monitoring CCPs	56%	47%
Corrective action	57%	55%
Verification	27%	25%
Documentation	40%	40%

the initiatives, and only 22% of businesses in both Wales and England indicated that they complied with current U.K. legislation by applying the first five HACCP principles. In both studies, businesses that were part of a larger group, and those that had food hygiene training provision in place were significantly more likely to have implemented a greater number of HACCP principles.

In both studies, approximately 90% of businesses sold cooked meat products in addition to raw meats. These businesses represent an increased risk to the consumer, as the potential for cross-contamination between raw and cooked product is greater. Of those, around 60% of businesses mainly sold cooked meats that were produced on the premises, while 25% tended to buy most of their cooked meats from a supplier. Overall, around 70% of all businesses were producing some cooked meats, and virtually all used probes to monitor internal food temperature as a critical control. About 20% of businesses indicated that they would be prepared to adopt a policy of only selling pre-wrapped cooked meats to reduce food safety risks.

HACCP Costs

HACCP costing questionnaires were completed by 100 respondents in the Welsh study and 129 respondents in the English study. Respondents were asked to identify a range of direct costs and time spent on various

Table 8.6 Costs of Planning, Implementing and Running HACCP in Welsh Butchers (n=96)

	Mean Direct Cost (£s)	*Mean Time (hours)*	*Mean Total Cost (£s) (1 hour = £5.50)*
Planning and implementation	£629	47h	£890
Weekly running cost	£27	25h	£166
Changes in weekly costs	+ £15	+ 6h	+ £51

activities during the planning, implementation, and running of a HACCP program. They were also asked to estimate the change in weekly running costs as a result of HACCP. Time input was costed at £5.50 Sterling per hour. A total of five businesses were not included (four in the Welsh study and one in the English study), as their capital costs and lost sales through closure during modification of premises severely skewed the data, and mean values were an unfair reflection of overall trends. Summary data of the costs are shown in Tables 8.6 and 8.7 for the Welsh and English studies, respectively. There are no significant differences in cost data for the two studies. This suggests validity of the methodological approach adopted in this evaluation.

The main element of planning and implementation cost was capital spending on equipment, which accounted for over two thirds of direct costs. Associated staff training costs were the second greatest expense. Consultancy costs had a minimal impact, but it must be remembered that the initiatives provided a considerable degree of support in HACCP imple-

Table 8.7 Costs of Planning, Implementing and Running HACCP in English Butchers (n=128)

	Mean Direct Cost (£s)	*Mean Time (hours)*	*Mean Total Cost (£s) (1 hour = £5.50)*
Planning and implementation	£520	62h	£859
Weekly running cost	£30	26h	£172
Changes in weekly costs	+ £8	+ 5h	+ £38

mentation. Time input for planning and implementation was dominated by training and planning time. Overall costs using the £5.50 Sterling per hour multiplier were dominated by equipment, training, and planning, however, much of the time involved did not result in higher wage bills, suggesting that much of the planning and training took part in the normal working week or in the spare time of the business managers.

Direct running costs were dominated by increased cleaning costs and lost turnover due to product loss. On average, direct running costs represented 1% of weekly turnover. Running time was spent mainly on cleaning and HACCP-related activities of monitoring, recording, and documentation. Overall costs represented approximately 6% of weekly turnover per business. The cost of cleaning represented 3 to 4% of the average weekly turnover. The reported increase in product loss was probably due the introduction of more stringent control combined with failure to implement adequate corrective action procedures.

The overall change in direct weekly costs was due mainly to increased staff wage bills, although this did not affect all businesses. The lower average in the English study was due to 15% of the businesses that reported an overall gain in turnover, leading to an average gain of about £1.00 Sterling per week, compared to Welsh butchers who noted an average turnover loss of about £8.00 Sterling per week. No businesses noted any weekly time savings as a result of implementing HACCP. Changes in total costs were noted by 90% of businesses and ranged from a modest saving to additional costs exceeding £300 Sterling per week. The average cost change was approximately 1.5% plus of weekly turnover, but it must be noted that most of this increase is the cost of time, which appears to have been found from the normal working week or spare time of the business managers.

Managerial Attitudes and Beliefs Towards Food Safety Management

Managers were asked about their attitudes and beliefs towards food safety control, HACCP, and the accelerated HACCP initiative, including the costs and benefits to the business. Tables 8.8 and 8.9 show summary attitudinal scales for the Welsh and English studies, respectively.

There was strong agreement from both studies that HACCP would not have been possible without the help provided by the initiative (EHO in Wales, and MLC-appointed consultants in England; *statement 1*). However, opinion was divided on whether information on HACCP was widely available before the initiative (*statement 2*). Just under half the respondents in each study believed that they would not have implemented HACCP if the business had had to pay for the training and advice provided by the ini-

Table 8.8 Attitudes/Beliefs towards HACCP and HACCP Initiative in Welsh Butchers (n=129)

		(% of cases)				
		Strongly Agree	*Agree*	*Neither Agree or Disagree*	*Disagree*	*Strongly Disagree*
1	It would not have been possible to implement HACCP without the help of the EHO	37	42	12	8	2
2	Information on HACCP was widely available before the initiative	6	33	26	33	2
3	HACCP would not have been implemented if the business had to pay for the training and advice from the initiative	12	38	32	15	3
4	Food safety is under greater control since implementing HACCP	33	50	13	2	2
5	HACCP simply reflects good hygiene practice	36	60	3	1	-
6	HACCP has had little impact on daily running of the business	5	37	27	25	6
7	The business will not need future help to maintain and update the HACCP system	6	30	24	38	2

Table 8.8 ***(continued)***

		(% of cases)				
		Strongly Agree	Agree	Neither Agree or Disagree	Disagree	Strongly Disagree
8	HACCP reduces the risk of the business causing food poisoning	34	50	8	6	3
9	HACCP increases staff awareness of food hygiene issues	28	63	5	3	1
10	HACCP helps ensure compliance with UK legislation	22	67	10	1	-

tiative (*statement 3*). Around 80% of respondents agreed that food hygiene was under better control since implementing HACCP (*statement 4*), and an even higher proportion agreed that HACCP simply reflects good hygiene practice (*statement 5*). Opinion was divided over whether HACCP had had any impact on the daily running of the business (*statement 6*), 30 to 40% agreeing with statement and 30 to 40% disagreeing. Similarly, a roughly equal proportion of businesses agreed or disagreed that they would need further help in maintaining and updating HACCP (*statement 7*). There was, however, broad consensus that HACCP would reduce the risk of the business causing food poisoning (*statement 8*), that HACCP increased staff awareness of hygiene issues (*statement 9*), and helped to ensure compliance with U.K. legislation (*statement 10*).

Impact on Hygiene Management, Knowledge, and Ability

Respondents were asked to rate their perceived knowledge of HACCP and hygiene issues and their ability to implement and manage HACCP after training. They were also asked to retrospectively rate their perceived knowledge and ability before receiving training under the HACCP initiative. A 5-point scale was used, where 1 represented no knowledge/ability and 5 represented complete knowledge/ability. Table 8.10 summarizes the changes in perceived knowledge and ability for Welsh

Table 8.9 Attitudes/Beliefs towards HACCP and HACCP Initiative in English Butchers (n=191)

		(% of cases)				
		Strongly Agree	*Agree*	*Neither Agree or Disagree*	*Disagree*	*Strongly Disagree*
1	It would not have been possible to implement HACCP without the help of the MLC	27	46	13	10	3
2	Information on HACCP was widely available before the initiative	3	31	16	43	7
3	HACCP would not have been implemented if the business had to pay for the training and advice from the initiative	12	36	25	24	4
4	Food safety is under greater control since implementing HACCP	29	49	14	7	2
5	HACCP simply reflects good hygiene practice	18	63	7	11	1
6	HACCP has had little impact on daily running of the business	3	27	18	43	8
7	The business will not need future help to maintain and update the HACCP system	6	27	32	33	2

Table 8.9 (continued)

		(% of cases)			
		Strongly Agree	*Agree*	*Neither Agree or Disagree*	*Disagree*	*Strongly Disagree*
8	HACCP reduces the risk of the business causing food poisoning	28	55	10	5	2
9	HACCP increases staff awareness of food hygiene issues	18	74	5	3	-
10	HACCP helps ensure compliance with UK legislation	13	72	18	6	2

and English respondents. Respondents in both studies felt that they had moderate knowledge of general food hygiene issues, and only a little knowledge of HACCP pre-training. Managers in businesses that operated a greater number of HACCP principles and prerequisite programs before training gave significantly higher pre-training ratings. Average post-training scores indicated that respondents' knowledge had increased to a level described as "much knowledge". Significant improvements were seen in knowledge of HACCP, and more specifically HACCP record-keeping and documentation systems.

Table 8.10 Perceived Knowledge of Hygiene and HACCP and Ability to Carry Out HACCP Tasks before and after Training under the Accelerated Initiative

	mean knowledge/ability rating			
	Welsh Study (n = 129)		*English Study (n = 191)*	
1 = no knowledge/ability 5 = complete knowledge ability	*Before Training*	*After Training*	*Before Training*	*After Training*
1 Hygiene knowledge	3.4	4.3	3.4	4.3
2 HACCP knowledge	2.7	4.3	2.7	4.3
3 HACCP ability	3.2	4.4	2.7	4.4

There was a slight difference in pre-training HACCP ability ratings between the two studies, but post-training ratings were the same for both Welsh and English businesses. The increase in ability rating was significant in the English study. Again, businesses that had more HACCP principles in place before training had a significantly higher pre-training ability rating.

Discussion

This study represents a unique multidisciplinary collaboration between academics, industry, and government and non-government organizations to evaluate the impact of accelerated HACCP initiatives in the retail butcher sectors in Wales and England. The two HACCP initiatives adopted quite different strategies and employed different methods, with the Welsh initiative using national government bodies (The Food Safety Technical Panel and local environmental health departments). Environmental Health Officers who act as enforcement officers provided training and consultancy. The English initiative was organized through the Department of Health and managed by the Meat and Livestock Commission, a non-departmental government organization whose primary role is marketing of the meat industry. A number of individuals, both private and from educational institutes, provided training and consultancy. This study did not evaluate the mechanism or content of the HACCP training, but examined the economic impact that training had on the recipients and the effect on hygiene and HACCP knowledge, attitudes, and practice.

It is important to recognize that the study used self-reporting through a postal questionnaire, and no attempt was made to validate the responses through an audit process. The collection of economic impact data in particular is difficult using this method, but the strategy was adopted in an attempt to get quantitative data from a comparatively large sample of the population. While individual case studies would perhaps have given more reliable data, it would have been impossible to extrapolate findings to the population. The methods used have previously been evaluated through a case-study approach (Mortlock et al., 2000b). Operational differences between the two initiatives and experience from evaluation of the Welsh study resulted in the use of different methods in evaluation of the English initiative. Fears that this may have resulted in some bias are resolved to an extent by the similarity of findings from both studies.

The results presented highlight the value of quantitative evaluations of programs such as the accelerated HACCP initiative. The use of questionnaires, attitude scales, and costing estimates provided a wealth of data that has allowed an assessment of the impact that introducing HACCP has had on the industry, and allows some insight into the future effect on regulatory authorities and the consumer. They also provide significant aid in the development of future strategies for implementing HACCP in other sectors of the industry.

The use of a purpose-made database to collate and analyze costing data is another unique development. Developed initially for case-study analysis, its use to collect data from over 200 butcher shops across Wales and England has validated the methodological approach to obtaining quantitative economic impact data. Given further development, this could form the basis of a predictive tool for use with other industry sectors.

HACCP costs were largely within the financial means of the businesses, although the largest strain was on the smallest businesses. The HACCP initiatives provided free training and consultancy, without which the costs of implementation would have increased. Many businesses would have found this a barrier to implementation. (Mortlock et al., 1999)

Businesses were generally not knowledgeable about HACCP prior to the initiatives, although the small number of businesses who had implemented HACCP components had a significantly greater knowledge. Previous research has shown that lack of awareness and the difficulty of obtaining useful information present barriers to HACCP use (Mortlock et al., 1999). The initiatives overcame both the training and awareness/technical support barriers. Small businesses are most likely not to perceive a need for formal hygiene management systems (Mortlock et al., 1999), and the wide agreement in this study with the statements that HACCP reduced the risk of the business causing food poisoning and increased staff hygiene awareness are indicators of the overall success of the training programs.

The impact of HACCP on weekly running costs is also within the means of most businesses. Increases are relatively small and focus mainly on improved cleaning procedures, CCP monitoring, and documentation. Much of the increase was in time commitment and did not lead to higher wage bills, suggesting that businesses are likely to be busier as a result of HACCP implementation or will need to rationalize current activities.

HACCP must be participatory, and any external consultancy or training must provide a business with the necessary skills and tools needed to be self-sufficient and implement, run, and review its own HACCP program. Over 30% of businesses indicated the need for future assistance, suggesting that the initiative provided solutions but did not develop HACCP independence.

Impact on Food Safety

The data suggest that the HACCP initiatives gave business managers a greater perceived ability to implement and manage all seven principles of

HACCP. Despite many managers' attitudes that they will need future assistance with HACCP, the fact that they have a significantly greater awareness of and positive attitude towards HACCP provides some evidence that food safety risks should be reduced.

The development of HACCP plans necessitated a detailed examination of existing prerequisite programs, and it is interesting to note that the greatest cost involved in running HACCP was the increase in cleaning costs. The prevention of cross-contamination in many small businesses relies not only on physical separation of cooked and raw meats, but also temporal separation and thorough cleaning between different activities. Increased awareness and hence improved cleaning represents a major impact on food safety risks. Similarly, the routine monitoring of temperatures within cold stores/display cabinets increased with HACCP use, with a potential positive impact on food safety.

It was also clear that the majority of managers believed that the introduction of HACCP had increased their food handlers' awareness of hygiene issues. Whether this results in better hygiene practice was not tested, and the Food Safety Research Group is involved in the development of robust tools to assess the impact of training on hygiene behavior.

The implementation of HACCP and proper management of prerequisite programs has helped the industry comply with current food hygiene regulations. Regulations were far from universally applied prior to the formal implementation of HACCP.

The Pennington Report called for a licensing system for retail butchers (Pennington, 1997), and there are plans to introduce such a scheme during 2000. A license would be dependent on implementation of an adequate HACCP program, and the initiatives are likely to lead to a smoother and less costly transition to licensing.

A properly documented HACCP system will also aid businesses and local authorities during routine audits and risk-based assessments of premises and practices by enforcement officers.

Any initiative designed to improve food safety control in an industry sector requires evaluation. The retail butchery sector has a high proportion of small businesses with owner/managers. It is recognized that such businesses experience difficulties in implementing HACCP. The initiatives in Wales and England appear to have positively affected business food hygiene management systems. Though the individual costs may be difficult to determine and the benefits may not be tangible to the individual businesses, the impact on both society and government could be significant.

Lessons Learned and Future Research Needs

The authors identified the need for evaluation of two accelerated HACCP initiatives and approached the body responsible for overseeing the Welsh initiative. Until that point there had been no plan for any analysis of the plan's impact. Agreement on the methods and process of evaluation took approximately one year, by which time the initiative had been implemented. Similarly, in the English study the authors raised the issue of evaluation, and entered a period of discussion over the methods and process that would be followed. Because of the lack of any coherent evaluation strategy and protracted discussions of the methods, no baseline data were gathered from the sector prior to the intervention.

Evaluating the HACCP systems that resulted from the initiatives was never an aim of this study. Nevertheless, ongoing longitudinal studies will assess whether the investment in developing and delivering the HACCP initiatives results in any long-term tangible benefits. One approach to this would be to examine food poisoning outbreak data in relation to retail butchers, but epidemiological studies are difficult, particularly given the number of sporadic cases associated with some foodborne pathogens. The maintenance and review of the HACCP systems should also be examined, particularly as some businesses may have introduced HACCP simply because it was expedient to do so in light of the rumored licensing system. Given this, management commitment to HACCP must be tested over time. One of the problems with a nationwide initiative is that appropriate methods must be developed and evaluated to ensure that a reliable indication of practices, knowledge, and commitment can be followed over a period of time. The use of a simple on-line database to gather cost data from a selection of businesses, and internet-based transmission of data for analysis is one such method that requires development.

The evaluation of HACCP, particularly if it is a criterion for licensing, requires a robust generic tool to enable objective auditing. Enforcement of any licensing regulations must be seen to be uniform and fair, particularly in Wales, where the local authority has a vested interest in the success of butchers' HACCP, having been directly involved in their implementation.

The data presented in this study show positive effects on management attitudes and self-reported knowledge and practices. A multidisciplinary approach to food safety research questions whether such positive effects are translated into improvements in practice. The Food Safety Research Group at UWIC is focusing on food handlers' behavior in a range of collaborative studies involving microbiologists, social scientists, consumer scientists, and psychologists. The aims of the research include determining the effect of a range of intervention strategies on consumer behavior in the domestic kitchen, and in turn how this effects the microbiological safety of the food. Observation and analysis tools have been developed for this research, and these are now being used to examine the psychological constructs that determine professional food handlers' behavior. This is being used to develop social cognition models that may explain and even predict behavior and the effectiveness of particular training strategies in changing behavior. Another study is using observational tools and microbiological and ATP bioluminescence hygiene testing to examine the extent and significance of cross-contamination events in catering establishments.

This ongoing research is essential if the full potential of other HACCP initiatives is to be realized. The catering sector is an important potential target for future initiatives, as it presents a risk to food safety (Griffith, 2000) and is dominated by a large number of small businesses (Mortlock et al., 1999). The evaluation of the butchery initiatives together with a better understanding of food hygiene behavior will identify the most appropriate sources and channels of information, effective learning and teaching strategies, factors that motivate business managers, and perceived barriers to HACCP. This will allow the development of better educational and training tools.

Another factor that will affect the success of HACCP implementation is the organizational culture of a particular business. Ensuring good hygiene behavior requires the right system plus the right management culture. This should show commitment from the top, and a management framework for the implementation of good hygiene practice. HACCP systems should be a working part of management practice, and not just manuals on a shelf.

Evaluating the impact of food safety control must be seen as an essential component of any initiative to effect change. It is imperative that the appropriate research-based methods are agreed upon before any implementation, and that baseline data is gathered to allow a direct before-and-after comparison of practices, costs, attitudes, and behavior. A multidisciplinary approach is the only way to effect this goal.

Acknowledgments

The authors thank the Welsh Food Safety Technical Panel, Meat and Livestock Commission, and U.K. Department of Health for facilitating the research.

References

Clayton, D.A., Griffith, C.J., Peters, A.C., and Price, P. (2000) Attitudes, knowledge and food safety practices of UK consumers, in: *Proceedings of the Second NSF Conference on Food Safety,* NSF International, Ann Arbor, MI, 321-328.

Coleman, P.D. and Griffith, C.J. (1998) Risk assessment - a diagnostic self assessment tool for caterers, *International Journal of Hospitality Management,* 17, 289-301.

Davidson, C.A., Griffith, C.J., Peters, A.C., and Fielding, L.M. (1999) An evaluation of two methods for monitoring surface cleanliness - ATP bioluminescence and traditional hygiene swabbing, *Luminescence,* 13, 1-5.

Griffith, C.J., Worsfold, D., and Mitchell, R. (1998) Food preparation, risk communication, and the consumer, *Food Control,* 9, 225-232.

Griffith, C.J. (2000) Food safety in catering establishments, in: *Safe Handling of Foods*, J.M. Farber and E.C.D. Todd, Eds., Marcel Dekker, New York, 235-255.

Griffith, C.J., Peters, A.C., Lewis, A., Davidson, C., Redmond, E. and Davies, C. (1999). The Application of Notational Analysis and Hazard Analysis to Assess Cross Contamination in Domestic Food Preparation, Department of Health Project Report DH215, London.

Harrison, W.A., Peters, A.C., and Fielding L.M. (2000) Growth of *Listeria monocytogenes* and *Yersinia enterocolitica* colonies under modified atmospheres at 4 and 8 °C using a model food system, *Journal of Applied Microbiology,* 88, 38-43.

Howes, M., McEwen, S., Griffiths, M., and Harris, L. (1996) Food handler certification by home study: measuring changes in knowledge and behaviour, *Dairy Food and Environmental Sanitation,* 19: 737-744.

Moore, G., Griffith, C.J., and Peters, A.C. (2000), Bactericidal properties of ozone, *Journal of Food Protection*, 63, 1100-1106

Mortlock, M.P., Griffith, C.J., and Peters, A.C. (2000a) A national survey of food hygiene training and qualification levels in the UK food industry, *International Journal of Environmental Health Research,* 10, 111-123.

Mortlock, M.P., Peters, A.C., and Griffith, C.J. (2000b) Applying HACCP to small retailers and caterers, a cost-benefit approach in: *Economics of HACCP: Studies of Costs and Benefits,* L.J. Innevehr, Ed., Eagan Press, St. Paul, MN, 301-314.

Mortlock, M.P., Peters, A.C., and Griffith, C.J. (1999) Food hygiene and Hazard Analysis Critical Control Point in the United Kingdom: practices, perceptions and attitudes, *Journal of Food Protection,* 62, 786-792.

Office of National Statistics (1998) New earnings survey 1998, H.M.S.O., London, Pennington Group (1997). Report on the circumstances leading to the 1996 outbreak of infection with *E. coli* O157 in central Scotland, the implications for food safety and the lessons to be learned, H.M.S.O., Edinburgh.

Peters, A.C., Griffith, C.J. and Mortlock, M.P. (2000) Evaluation of barriers to the use of food hygiene management systems throughout the UK food industry, Ministry of Agriculture, Fisheries, and Food Report FS1050, London.

Redmond, E.C., Griffith, C.J., Price P., and Peters, A.C. (1999) Traditional and social market approaches to consumer food safety education, in: *Proceedings of the XIX International Consumer Studies and Home Economics Research Conference,* C.J. Strugnell and G.A. Armstrong, Eds., University of Ulster, Belfast, 126-127.

Various authors (1999) Food Microbiology and Food Safety into the Next Millennium, A.C.J. Tinjtelaans, R.A. Samson, F.M. Rambouts, and S. Notermans, Eds., conference proceedings, Foundation for Food Micro '99 c/o TNO, The Netherlands.

Woollen, A. (1999), Safety and the Y2K, *Food Processing,* February: p.20.

Chapter 9

Assessing the Bases of Food Safety Concerns

Rex H. Warland, Robert O. Herrmann, and Arthur Sterngold

Over the last decade, the American public has become increasingly concerned about food safety. Studies from a variety of social science disciplines have found a wide range of variables related to food safety concerns. In this chapter, all these variables are brought together in a single study to investigate the bases of these concerns regarding the safety of food. A national telephone survey of 1400 adults conducted in 1999 provided the data to examine this more comprehensive model. Results show a variety of factors from all the disciplines considered related to food safety concerns. The data indicate that there are multiple bases of food safety concerns, which are more complex and diffuse than previous studies have suggested.

Introduction

Food safety concerns have become a major issue for the American public. As increasing numbers of Americans have become concerned about food-related hazards such as *Salmonella* and *Escherichia coli,* social scientists from a variety of disciplines have conducted studies focused on identifying the factors related to food safety concerns. A majority of these studies have investigated the demographic characteristics of those who have con-

0-8493-2217-0/01/$0.00+$.50

cerns about food safety. Other studies have considered factors such as consumers' perceived control and knowledge of food risks, information seeking related to food safety, and the degree to which the public trusts the government and other components of the food system to provide consumers with safe food products.

The joint relationship of all these factors to food safety concerns is explored here. The data presented were drawn from a national telephone survey of 1400 adults conducted in mid 1999. These data provide a much more comprehensive profile than previous studies of factors that underlie food safety concerns. A new approach to measuring food safety concerns will also be demonstrated.

Review of Past Studies

Previous studies of factors related to food safety concerns can be classified into one of three broad categories. The first category includes studies of the demographic characteristics of those concerned about food-related hazards. A second group of studies has dealt with individual experiences with food safety and with respondents' perceptions of their vulnerability to, control over, knowledge of, and involvement with food-related hazards. The third set of studies has focused on the American public's confidence in the food system to produce safe food products. The findings from these three sets of studies will be summarized to provide a guide for the construction of our conceptual model.

Demographic Characteristics

During the last decade, several dozen studies have investigated the demographic profile of those concerned about food safety. These studies have examined demographic characteristics of those concerned about pesticides (Dunlap and Beus, 1992; Flynn et al., 1994), recombinant bovine somatotropin (bST) (Grobe and Douthitt, 1995), genetic engineering (Hoban et al.,1992), antibiotics (Nayga, 1996), and general food safety (Jordan and Elnagheeb, 1991; Lin, 1995).

The most consistent finding is that women express greater concern than men about food safety. Bord and O'Connor (1997) have reviewed a wide range of studies related to environmental concerns, food safety concerns, health concerns, and crime fears. Women consistently expressed more concern than men. Bord and O'Connor (1997) attributed this finding to the likelihood that women believe they are more vulnerable to risks than men. Flynn et al. (1994) compared the concerns of men and women for about 25 different risks, including pesticides, bacteria in food, food irradiation, and genetic engineering. These researchers found that white women had higher mean risk ratings than white men for all 25 hazards. Flynn et al. (1994) suggest that women may be more concerned because of different risk experiences, different socialization experiences, and less willingness to accept levels of risk advocated by experts.

Gender differences have consistently been reported by studies which have focused on concerns for food-related hazards. For example, Dunlap and Beus (1992), Grobe and Douthitt (1995), Hoban et al. (1992), Lin (1995), and McGuirk et al. (1990) found that women were more concerned about a variety of food-related hazards.

Another consistent finding is that non-whites, lower educated, and lower income respondents are more concerned about food safety. Beck (1992) has hypothesized that an individual's class position in society is related to how vulnerable he is to risks. These groups are not only more exposed to health risks, but have less information and resources to protect themselves. Beck (1992) indicates that education and attention to information reduces risks associated with nutrition and food safety.

On the basis of Beck's (1992) hypothesis, it would be expected that those who are most vulnerable would be most concerned about food-related hazards. Past studies of food safety concerns support this hypothesis. Flynn et al. (1994), Nayga (1996), McGuirk et al. (1990), and Herrmann et al. (2000), among others, have found that non-whites were more concerned about food safety issues than whites. Nayga (1996), Jordan and Elnagheeb (1991), and Grobe et al. (1995) report lower-educated respondents were more concerned about food safety than better-educated respondents. Lower income was found to be associated with food safety concerns by Dunlap and Beus, (1992), Grobe and Douthitt (1995), Hoban et al. (1992), Nayga (1996), Jordan and Elnagheeb (1991), and Herrmann et al. (2000).

Two other demographic characteristics have been included in studies of food safety concerns. Although the results have not been as consistent as those for gender, race, education, and income, these studies have found that respondents who are older are more likely to be concerned about food-related hazards than those who are younger (Nayga, 1996; Jordan and Elnagheeb, 1991; Lin, 1995; and Herrmann et al., 2000). A few studies have investigated the relationship of presence of children to food safety concerns. McGuirk et al. (1990) and Herrmann et al. (2000) found that those respondents who had younger children were most concerned about food safety, but Nayga (1996) found no relationship between concern and presence of children.

The general pattern that emerges from this brief review is that women, non-whites, lower educated, lower income, older persons, and those who have young children, are most concerned about food safety issues. While statistically significant, the collective effect of these demographic variables is modest. The R^2 values range from three to eight percent of the total variance. It is clear that factors other than demographics need to be included in a model of factors related to food safety concerns.

Perceived Vulnerability and Involvement

The second category of studies has investigated individuals' experiences and perceptions of food safety. This category includes studies of experience with food poisoning, the extent to which individuals believe they are vulnerable to becoming ill from unsafe food, perceived knowledge about food-related hazards, perceived control over exposure to food-related hazards, and involvement with food safety issues. We will consider the studies related to these factors next.

Fein et al. (1995) have reported that people who believe they have experienced a foodborne illness were more concerned about food safety issues than those who did not report experiences with foodborne illness. They speculate that a recent foodborne illness experience may sensitize the victims to food safety issues. A related factor is individuals' perception of the likelihood they will become ill from unsafe food. Beck's (1992) theory suggests those who believe they are vulnerable to food-related hazards are also more likely to be concerned about food safety issues.

Perceived knowledge about food safety and perceived control over exposure to food-related hazards are factors that also need to be considered when investigating the bases of food safety concern. Food researchers have considered perceived control as a key factor when studying dietary behavior (AbuSabha and Achterberg, 1997) or food safety behavior (Frewer et al., 1994; Schafer et al., 1993). The results of the studies that have related perceived control to food safety concerns have been mixed. Grobe et al., (1999) report that a general feeling of lack of control over one's life was associated with increased concern about bST. Sparks and Shepherd (1994) found that respondents believed they had little control over exposure to food-related hazards such as *Salmonella*, genetic engineering, pesticides, and *Listeria,* all hazards which studies have shown are of concern to the public. On the other hand, Frewer et al. (1994) found no direct relationship between perceived control and perceived risk of food-related hazards.

The relationship between perceived knowledge of food safety and concern is also not clear. Bord and O'Connor (1997) found a positive relationship between self-assessed knowledge of environmental hazards and concern about these hazards. However, Frewer et al. (1994), Sapp et al. (1994), and Warland et al. (1999) found little relationship between knowledge and concern.

If concerns about food safety are strong and salient, then these issues should be important in the respondent's everyday activities and decision-making. According to Mason et al. (1988), if attitudes or perceptions are salient, those who hold these attitudes should think about the issues often and base behavioral decisions on their attitudes. Demerath (1993) has argued that the issues related to developed attitudes will frequently be discussed with others, and that an individual will think about the issues often if the set of cognitions that underlie the attitudes are developed and consistent. Grobe et al. (1999) have suggested that if individuals are concerned about food safety issues, they are more likely to search for information and be more attentive to risk communications. We will include measures of involvement in our model to determine if food safety concerns are indeed salient attitudes or less developed attitudes.

This review of hypotheses and empirical studies related to perceived vulnerability and involvement indicates that those most concerned about food-related hazards are also likely to have experience with foodborne illness, believe they are vulnerable to food-related hazards, and have salient attitudes about food safety issues. While some studies have found that perceived control over exposure to food-related hazards and perceived knowledge of these hazards were related to concerns, other studies have found no relationships between these factors and concern. The study reported here will revisit the relationship between perceived knowledge and concern, and between perceived control and concern, to investigate whether these relationships exist.

Confidence in the Food System

Perceptions of risks are often a function of the degree to which the institutions responsible for the assessment and management of risk are trusted (Short, 1984). Frewer and Shepherd (1994), among others, have pointed out that risk is socially constructed. They suggest that the best predictors of risk are not demographic characteristics, but the degree to which people trust those who manage risks.

The few studies that have related trust of the food system to concerns about food safety have consistently found a negative relationship. Lower levels of trust are related to higher levels of concern. Dunlap and Beus (1992) found that those who did not trust the food industry's use of pesticides also were concerned about the safety of pesticides. Jussaume and Higgins (1998) reached a similar conclusion, finding that lack of faith in the government's role in ensuring food safety was strongly associated with food safety concern. Both Sapp et al. (1994) and Dittus and Hillers (1993) found a negative relationship between trust and concern.

Our model will include a measure of public trust of the major components of the food system. This approach will make possible a more comprehensive test of the relationship between trust and concern.

Methods

The data for this study were collected in a nationwide telephone survey conducted in June and July 1999. Random-digit dialing (RDD) procedures were used to reach both listed and unlisted telephone numbers throughout the 48 contiguous states. A total of 1400 adults (18 years or older) were interviewed. In order to obtain roughly equal numbers of male and female respondents, interviewers first asked to speak to an adult male. If one was not available, the interviewer asked to speak to a female. As a result, 45% of the respondents were men. The cooperation rate, i.e., the percentage of contacts eligible to be interviewed completing an interview, was 59%.

The sample was representative of the adult U.S. population with respect to race, marital status, age, household size, and employment status. Those who had a high school diploma or some college or technical training were appropriately represented, those with less than a high school diploma were underrepresented, and those who had completed college, graduate school, or professional school after college were overrepresented.

The dependent variables for this study included concern about pesticides, *Salmonella,* and *E. coli.* Over the last decade, we have developed a new method of measuring concern. In the past, the most common formulation for a concern question was "how concerned are you about . . . . ?" This question form is very common in surveys conducted by government agencies, universities, and private marketing firms (Herrmann et al., 1998). Few have recognized the problem created by the presupposition embedded in asking "how concerned are you about . . . ?" Studies of the social use of language indicate that this question form contains unstated assumptions about the respondents (Sterngold et al., 1994). The wording of the question may give the respondent the impression that he should be concerned. Respondents may feel they should conform to these presuppositions even when it is inappropriate to do so. When respondents agree to be interviewed, they also agree to go along with the "rules of the game" imposed by the interviewer and question formats (Herrmann et al., 1998).

In a series of studies conducted in 1991, 1994, 1995, and 1997, we developed an alternative approach to measuring concern by using a concern filter. Employing a split-ballot method, those interviewed were randomly assigned to one of two groups. One group was asked the questions in the conventional form, i.e., "how concerned are you about . . . ?" The other group was given an initial concern filter question which first asked "are you concerned about . . . ?" Those who indicated they were concerned then were asked how concerned they were in a follow-up question (Sterngold et al., 1994). The major effect of the concern filter was to produce a lower percentage of those who indicated some degree of concern and to increase the percentage of those who expressed no concern. Over the 25 split-ballot experiments conducted from 1991 through

1997, the percentage of respondents expressing no concern with the filter form was on average double that obtained with the conventional form. For example, 14% said they were "not concerned" about giving hormones to dairy cows using the conventional question, but 32% expressed no concern using the filter (Sterngold et al., 1994).

For this study, all respondents were questioned about their food safety concerns using the filter form. Those interviewed were first asked, "Are you concerned about pesticides and chemical residues on the fruits and vegetables you eat?" The response categories were "Yes," "No," or, "Never heard of pesticides." If the respondent answered, "Yes," they were then asked, "Are you very concerned, somewhat concerned, or a little concerned?" The question format was the same for *Salmonella* and *E. coli.* "Very concerned" was scored "4;" "somewhat concerned" was scored "3;" "a little concerned" was scored "2;" and "not concerned" was scored "1."

The frequency distribution of three dependent variables is displayed in Table 9.1. Those interviewed typically expressed concern about pesticides, *Salmonella,* and *E. coli.* Over 70% of those interviewed were "very concerned" or "somewhat concerned" about these three food-related hazards. The data also indicate the American public is quite familiar with these food safety risks. Less than 2% of those interviewed had not heard of pesticides or *Salmonella,* and only 6% had never heard of *E. coli.*

Table 9.2 summarizes the measures of the independent variables. Our model includes six demographic variables: gender, education, age, presence of children age 10 or under, role in food preparation, and race. Income was not included because 1) preliminary analysis found that income was not related to concern about any of the three food-related hazards, and 2) a large number of cases (324) would have been lost because respondents either did know their household income or refused to answer the income question. Role in food preparation was included in this analysis to determine if those who did most or all of the food preparation were more or less likely to be concerned about food-related hazards than those who had less food preparation responsibility.

Table 9.1 Frequency Distribution of Concern about Pesticides, *Salmonella,* and *E. coli*

Responses	*Pesticides (%)*	*Salmonella (%)*	*E. coli (%)*
Very concerned	42.1	44.5	41.5
Somewhat concerned	33.4	30.5	28.7
A little concerned	3.8	5.2	4.9
Not concerned	20.3	18.5	18.9
Never heard of it	0.4	1.3	6.0

Table 9.2 Measures of the Independent Variables

Independent Variables	*Measure*
Demographic Characteristics	
Gender	Male = 1; female = 0
Education	Grade school = 1; some high school = 2; completed high school = 3; some college or technical school = 4; completed college = 5; graduate or professional school = 6
Age	Age in years
Children age 10 or under present	Yes = 1; no = 0
Involvement in food preparation	Do hardly any or none = 1; some or share = 2; do most = 3; do all = 4
Race	White = 1; African American = 2; Hispanic = 3; other non-white = 4
Perceived Vulnerability and Involvement	
How likely to get sick from unsafe food	Not at all likely = 1; not very likely = 2; somewhat likely = 3; very likely = 4
Perceived knowledge of (pesticides, *Salmonella, E. coli*)	Know nothing = 1; know a little = 2; know some = 3; know a great deal = 4
Perceived control over amount of (pesticides, *Salmonella, E. coli*)	Have no control = 1; have a little control = 2; have some control = 3; have a great deal of control = 4
Recent experience with food poisoning	Yes, had food poisoning in last 12 months = 1; no = 0
Avoid certain foods because likely to be unsafe	Yes = 1; no = 0
Attention to news on how to keep food safe	Pay not much attention = 1; pay some attention = 2; pay a lot of attention = 3
Attention to news about food safety scares or product recalls	Pay not much attention = 1; pay some attention = 2; pay a lot of attention = 3
How frequently think about food safety	Hardly at all = 1; once in a while = 2; several times a week = 3; everyday = 4
Think about food safety last time shopped for food	Yes = 1; no = 2; don't shop for food = 3

Table 9.2 *(continued)*

Independent Variables	*Measure*
Confidence in Food System	
Confidence in food system index	An index created from responses to statements asking how much confidence (very, somewhat, not sure, not very confident) respondents had in the safety of food imported, sold in restaurants and supermarkets, produced by farmers and food processors, and checked by government inspectors
Need for government inspectors	Fewer = 1; no change = 2; not sure = 3; more = 4

Our model includes nine measures of perceived vulnerability and involvement. The measures include questions about whether they or anyone in their household had food poisoning in the last 12 months, the likelihood they would get sick from unsafe food, and whether they avoided foods they believed to be unsafe. Those interviewed were asked how much they knew about pesticides, *Salmonella,* and *E. coli,* and how much control they believed they had over the amount of these three food-related hazards in the foods they eat. There were four measures of involvement, including how often those interviewed thought about food safety, whether they thought about food safety the last time they shopped for food, and the amount of attention they paid to news about how to keep food safe and news about food safety scares and recalls.

The third set of independent variables included two measures of the respondents' confidence in the food system. The first measure was an index consisting of six measures of confidence in the safety of food that is imported, sold in restaurants and supermarkets, produced by farmers, produced by food processors, and the job government inspectors do to make sure our food is safe. The Cronbach's coefficient alpha for the index was 0.76 which is in the recommended range (Peterson, 1994). The second measure was whether those interviewed believed we need more, fewer, or the same number of government food inspectors.

Results

The regression coefficients estimating the relationships of pesticide concerns to the demographic characteristics, the perceived vulnerability, and involvement variables, and confidence in the food system measures are presented in Table 9.3. The demographic profile (Model A) is very similar to the findings of the earlier studies reviewed in the first part of this chapter. Women, lower educated, older, and non-white respondents were the most concerned about pesticides. The predictive power of this demographic model was low. The R^2 value is only 0.05.

Seven of the nine perceived vulnerability, and involvement variables (Model B) were significantly related to concern about pesticides. All of the involvement measures related to pesticide concern. Those who thought

Table 9.3 Regression Coefficients for Variables Associated With Concern About Pesticides

Independent Variables	*Model A N = 1321*	*Model B N = 1314*	*Model C N = 1393*	*Model D N = 1248*
Demographic Characteristics				
Gender (male = 1, female = 0)	−.26***			−.04
Education	−.09***			−.08***
Age	.01**			.00
Children age 10 or under	−.05			−.15*
Involvement in food preparation (most, all = 1, other = 0)	.01			−.10
African American[a]	.42***			.24**
Hispanic[a]	.19			.11
Other non-white[a]	.42***			.25*
Perceived Vulnerability and Involvement				
Believe likely to get sick from unsafe food		.02*		.01
Perceived knowledge of pesticides		.08*		.06
Perceived control over amount of pesticides		.05		.08**
Recent experience with food poisoning (yes = 1, no = 0)		−.09		−.06
Avoid certain foods because likely to be unsafe (yes = 1, no = 0)		.20***		.11
Attention to news on how to keep food safe		.18***		.17**
Attention to news about food safety scares		.22***		.18**

Table 9.3 *(continued)*

Independent Variables	*Model A* N = 1321	*Model B* N = 1314	*Model C* N = 1393	*Model D* N = 1248
Frequently think about food safety		.20***		.17***
Think about food safety when last shopped for food[b]		.37***		.35***
Don't shop for food[b]		.18		.15
Confidence in the Food System				
Confidence in food system index			−.09***	.07***
Need more government inspectors[c]			.37***	.21**
Not sure need more government inspectors[c]			.08	.21
Intercept	3.14***	.62**	4.06***	2.17***
Adjusted R^2	.05***	.17***	.10***	.23

[a]Masked category "white"
[b]Masked category "no"—did not think about food safety when last shopped for food
[c]Masked category "fewer" or "no change" in number of government inspectors
*p<.05, **p<.01, ***p<.001

about food safety issues, who thought about food safety when they last shopped for food, and paid attention to news about food safety scares and news about how to keep food safe were the most concerned about pesticides. In addition, those who avoided certain foods they believed to be unsafe, those who believed they were likely to get sick from unsafe food, and those who indicated they knew something about pesticides were also the most concerned. The R^2 for Model B is 0.17.

Both measures of confidence in the food system (Model C) were related to concern about pesticides. Those who have less confidence in the components of the food system and those who indicated more government inspectors were needed were the most concerned about pesticides. These two variables explained 10% of the variance.

The composite Model D included all the independent variables from Models A, B, and C. This combination of variables explained 23% of the variance, indicating that the three sets of variables together explain more variance than any individual set. Those with less education and those who were nonwhite were concerned about pesticides, but gender and age were no longer associated with concern. The presence of children was related to concern, with those who have children age 10 or under indicating less concern about pesticides.

Six of the variables which were significant in Models B and C were also significant in Model D. Perceived control was positively related to concern, but perceived knowledge, avoiding certain foods, and perceived vulnerability to getting sick from unsafe food were not significant. Overall, 10 of the 17 variables were significantly related to pesticide concern. This finding suggests pesticide concerns have multiple bases.

Table 9.4 displays the results of the regression analyses of the factors associated with concern about *Salmonella*. In general, the pattern of significant relationships is similar to that for the pesticides concern model. Women, older respondents, and non-white respondents were most concerned about *Salmonella* (Model A). Unlike the pesticide models, those with children age 10 or under were more concerned about *Salmonella*. Involvement in food preparation was related to concerns about this food-

Table 9.4 Regression Coefficients for Variables Associated with Concern about *Salmonella*

Independent Variables	*Model A N=1309*	*Model B N=1296*	*Model C N=1381*	*Model D N=1229*
Demographic Characteristics				
Gender (male = 1, female = 0)	−.37***			−.11
Education	−.04			−.07**
Age	.01**			.01*
Children age 10 or under	.18**			.02
Involvement in food preparation (most, all = 1, other = 0)	−.20**			−.29***
African American[a]	.41***			.26**
Hispanic[a]	.22			.16
Other non-white[a]	.20			.09
Perceived Vulnerability and Involvement				
Believe likely to get sick from unsafe food		.12***		.11***
Perceived knowledge of *Salmonella*		.16***		.18***
Perceived control over amount of *Salmonella*		−.03		−.01
Recent experience with food poisoning (yes = 1, no = 0)		.01		.05
Avoid certain foods because likely to be unsafe (yes = 1, no = 0)		.17**		.14*

Table 9.4 *(continued)*

Independent Variables	*Model A* *N=1309*	*Model B* *N=1296*	*Model C* *N=1381*	*Model D* *N=1229*
Attention to news on how to keep food safe		.39***		.37***
Attention to news about food safety scares		.22***		.21***
Frequently think about food safety		.15***		.11***
Think about food safety when last shopped for food[b]		.29***		.26***
Don't shop for food[b]		.20*		.09
Confidence in the Food System				
Confidence in food system index			−.05***	−.02**
Need more government inspectors[c]			.49***	.31***
Not sure need more government inspectors[c]			−.03	.13
Intercept	3.08***	.06	3.43***	.55
Adjusted R^2	.05***	.22***	.07***	.27***

[a]Masked category "white"
[b]Masked category "no" – did not think about food safety when last shopped for food
[c]Masked category "fewer" or "no change" in number of government inspectors
*p<.05, **p<.01, ***p<.001

related hazard, although this factor was not related to pesticide concern. Those who have major responsibility for food preparation were less concerned about *Salmonella*. The R^2 value remained low.

The pattern of relationships for the perceived vulnerability and involvement set (Model B) for *Salmonella* concern is very similar to that for pesticide concern. The only difference is that those who don't shop for food were more concerned about *Salmonella* than those who did not think about food safety when they last shopped for food. The same variables in Model C were significant for *Salmonella* concerns, as was the case for pesticide concerns.

In the composite Model D, eight of the variables that were related to pesticide concern are also related to *Salmonella* concern. Those who were less involved in food preparation, those who believed they were likely to get sick, those who knew more about *Salmonella,* and those who avoided certain foods were also more concerned about *Salmonella*. These variables were not related to pesticide concern in the composite model.

In summary, 13 of the 17 variables were related to concern about *Salmonella* and explained 27% of the variance. As was the case with concern about pesticides, a variety of factors were related to concern about *Salmonella*.

The variables associated with concern about *E. coli* are displayed in Table 9.5. With several differences, the variables related to *E. coli* concern

Table 9.5 Regression Coefficients for Variables Associated with Concern about *E. coli*

Independent Variables	*Model A N = 1249*	*Model B N = 1227*	*Model C N = 1315*	*Model D N = 1169*
Demographic Characteristics				
Gender (male = 1, female = 0)	−.31***			−.05
Education	−.03			−.04
Age	.01***			.00
Children age 10 or under	.14			.07
Involvement in food preparation (most, all = 1, other = 0)	−.06			−.15*
African American[a]	.33***			.16
Hispanic[a]	.37**			.29*
Other non-white[a]	.27			.10
Perceived Vulnerability and Involvement				
Believe likely to get sick from unsafe food		.08*		.07*
Perceived knowledge of *E. coli*		.12**		.15***
Perceived control over amount of *E. coli*		.07*		.07*
Recent experience with food poisoning (yes = 1, no = 0)		.01		.01
Avoid certain foods because likely to be unsafe (yes = 1, no = 0)		.21***		.16**
Attention to news on how to keep food safe		.28***		.25***
Attention to news about food safety scares		.29***		.31***
Frequently think about food safety		.19***		.17***
Think about food safety when last shopped for food[b]		.24***		.21**
Don't shop for food[b]		.23*		.19

Table 9.5 *(continued)*

Independent Variables	*Model A N = 1249*	*Model B N = 1227*	*Model C N = 1315*	*Model D N = 1169*
Confidence in the Food System				
Confidence in food system index			−.04***	−.02*
Need more government inspectors[c]			.44***	.28**
Not sure need more government inspectors[c]			.01	.06
Intercept	2.89***	−.03	3.38***	.17
Adjusted R^2	.04***	.21***	.05***	.25***

[a]Masked category "white"
[b]Masked category "no"—did not think about food safety when last shopped for food
[c]Masked category "fewer" or "no change" in number of government inspectors
*p<.05, **p<.01, ***p<.001

in Models A, B, and C are generally similar to those related to pesticide and *Salmonella* concern. In the composite Model D, 12 of the 17 variables were related to concern about *E. coli*. The only demographic variables that were related to *E. coli* concern were involvement in food preparation and being Hispanic. For the other two categories of variables, all variables except experience with food poisoning were significantly related to concern about *E. coli*.

The final regression analyses are shown in Table 9.6. The three dependent variables were combined into a concern index. The Cronbach's coefficient alpha for this index was 0.74, which is in the acceptable range (Peterson, 1994). Two other indexes were created, one for perceived knowledge of the three hazards and another for perceived control. The alpha coefficients were 0.68 and 0.67, respectively, values that are in the appropriate range.

The demographic characteristics (Model A) indicated that women, those with less education, those who are older, and those who are nonwhite are most concerned about these three food-related hazards. The R^2 value is 0.07.

The perceived vulnerability and involvement variables (Model B) were strongly related to the concern index. Those who believed they were vulnerable to getting sick from unsafe food, who indicated they knew something about the three food-related hazards, who avoided certain foods, thought about food safety, and paid attention to news stories about food safety scares and stories about how to keep food safe were most concerned. The R^2 value is 0.29, which is four times higher than the R^2 for the demographic characteristics. Both measures for the confidence in the food system (Model C) were also significantly related to the concern index.

Table 9.6 Regression Coefficient for Variables Associated with Food-Related Hazard Concern Index

Independent Variables	*Model A N=1237*	*Model B N=1184*	*Model C N=1303*	*Model D N=1127*
Demographic Characteristics				
Gender (male = 1, female = 0)	−.94***			−.29
Education	−.18**			−.18**
Age	.02***			.01*
Children age 10 or under	.27			−.08
Involvement in food preparation (most, all = 1, other = 0)	−.27			−.56***
African American[a]	1.14***			.68***
Hispanic[a]	.76*			.60*
Other non-white[a]	.88**			.53
Perceived Vulnerability and Involvement				
Believe likely to get sick from unsafe food		.19*		.16
Perceived knowledge index		.10**		.12**
Perceived control index		−.01		.01
Recent experience with food poisoning (yes = 1, no = 0)		−.08		−.01
Avoid certain foods because likely to be unsafe (yes = 1, no = 0)		.58***		.40**
Attention to news on how to keep food safe		.85***		.77***
Attention to news about food safety scares		.76***		.73***
Frequently think about food safety		.55***		.46***
Think about food safety when last shopped for food[b]		.95***		.89***
Don't shop for food[b]		.71**		.52*
Confidence in the Food System				
Confidence in food system index			−.18***	−.10***
Need more government inspectors[c]			1.26***	.79***

Table 9.6 *(continued)*

Independent Variables	*Model A* *N=1237*	*Model B* *N=1184*	*Model C* *N=1303*	*Model D* *N=1127*
Not sure need more government inspectors[c]			.07	.46
Intercept	9.06***	.95	10.89***	3.04***
Adjusted R^2	.07***	.29***	.11***	.36***

[a]Masked category "white"
[b]Masked category "no"—did not think about food safety when last shopped for food
[c]Masked category "fewer" or "no change" in number of government inspectors
*p<.05, **p<.01, ***p<.001

The composite Model D is the strongest model of those presented. The R^2 value is 0.36 and 12 of the 17 variables were significantly related to the concern index. All three groups of variables were related to the index. Lower educated, older, non-white respondents and those less involved in food preparation, those who reported they knew something about the three food risks, who avoided unsafe food, who thought about and read about food safety, and didn't trust the food system were the most concerned. It is clear that there are multiple sources of food safety concerns, and that these factors are additive, not redundant.

Discussion

Our purpose has been to provide a detailed profile of the factors that are related to food safety concerns. The regression analyses have uncovered a number of findings which should enrich our understanding of the bases of food safety concerns.

The models that only include demographic characteristics (Model A) identify the same groups that many earlier studies have found. Women, lower educated, older, and non-white individuals were the most concerned about food safety. These groups appear to be those most vulnerable and sensitive to risks in American society. However, the strength of the relationships to concern were weak, suggesting that these factors are not major predictors of food safety concerns.

Variables in the perceived vulnerability and involvement set (Model B) were, with the exception of recent experience with food poisoning, related to food safety concerns for one or more of the food hazards. The four measures of involvement were related to concern about all three food hazards. This finding suggests that concerns about pesticides, *Salmonella,* and *E. coli* are salient. In other words, those who were concerned about food safety issues paid attention to information, thought about these issues frequently, and based food purchase decisions, in part, on these concerns.

The other important finding in this set is the role of perceived knowledge and perceived control in food safety concerns. Perceived knowledge was positively related to concern about pesticides, *Salmonella, E. coli,* and the combined concern index. Perceived control related to concern about pesticides, and was only marginally related to concern about *E. coli*. These findings support earlier work that found weak or no relationships between perceived control and concern. However, these results indicate that those who indicated they knew about these food hazards were concerned about them, supporting the notion that food safety concerns are based on informed opinion.

The two measures of confidence in the food system (Model C) were related to concern in all four regression analyses. Those who were not confident of the safety performance of the food system and wanted more government involvement to ensure food safety were the most concerned about all three food hazards. These findings are consistent not only with findings in other food safety studies, but also findings in other areas of risk research including studies of environmental risks.

The composite models (Model D) present a new profile of those concerned about food safety. First, when all 17 variables are included, there are two major changes in the demographic profile. In all four composite models, gender was not significantly related to concern. This is surprising, since gender has been consistently related to concern in past studies. Bord and O'Connor (1997) did find, however, that when they constructed a composite model to explain concerns about environmental risks, gender was no longer significant when measures of perceived vulnerability were included. It may be that women feel more vulnerable to risks than men, so when control variables are introduced, the gender difference disappears.

Another new finding in the composite models is that those who have major responsibilities for food preparation were less concerned about *Salmonella, E. coli,* and the combination of these food-related hazards. We can only speculate about this finding, but it appears that those who were responsible for food preparation may have more confidence that they can control exposure to these hazards than those who have less experience in food preparation. We suggest this relationship be considered in future studies, because if others confirm this finding, the information should be useful to those in health and nutrition education.

The most important finding from these data analyses is that there are multiple bases for food safety concerns. In the four composite models, between 10 and 13 of the 17 relationships tested were statistically significant. Moreover, variables from all three sets of measures were statistically related to concern. There is also evidence that the three sets of variables each independently add variance in the composite model, i.e., none of the three models (Models A, B, and C) explained as much variance as the composite model (Model D). Clearly, the bases of food safety concern are much more complex and diffuse than previous studies have suggested.

The variables in the models presented here were drawn from a variety of disciplines including agricultural economics, rural sociology, sociology, psychology, social psychology, political science, nutrition, and marketing. Our approach has been to bring this diverse set of variables together to provide a more comprehensive profile of those who are concerned about food safety. The results indicate that variables from all these disciplines are related to food safety concerns. Moreover, the fact that a majority of the factors are related to concerns about food safety suggests that Americans currently do not have a shared set of perceptions about food safety. Rather, they have diffuse perceptions that are weakly linked to each other, and weakly linked to demographic characteristics as well.

The data presented here suggest that much more needs to be done to unravel the web of relationships that underlie concerns about food safety. We recommend that future studies use this larger set of variables from multiple disciplines, as well as promising variables from these and other disciplines. Our data indicate that the lower socio-economic groups, those who know about food safety, those involved with food safety information, and those who have reservations about the safety of our food system were the most concerned about food safety. These, however, are not the same groups of people. Rather, they are different segments of the American public. Over time, the perceptions of these various segments may become more organized and be based more closely on a single set of cognitions or possibly several different sets of perceptions which are connected to different demographic groups. We believe that these profiles are most likely to be discovered if the widest range of concepts from a number of disciplines is included in future empirical investigations.

References

AbuSabha, R., Achterberg, C., Review of self-efficacy and locus of control for nutrition- and health-related behavior, *Journal of the American Dietetic Association,* 97, 1122, 1997.

Beck, U., *Risk Society: Towards a New Modernity,* Sage, London, 1992, Chap. 1, 2.

Bord, R. J. and O'Connor, R. E., The gender gap in environmental attitudes: the case of perceived vulnerability to risk, *Social Science Quarterly,* 78, 830, 1997.

Demerath, L., Knowledge-based affect: cognitive origins of "good" and "bad," *Social Psychological Quarterly,* 56, 136, 1993.

Dittus, K. L. and Hillers, V. N., Consumer trust and behavior related to pesticides, *Food Technology,* 47, 87, 1993.

Dunlap, R. E. and Beus, C. E., Understanding public concerns about pesticides: an empirical examination, *Journal of Consumer Affairs,* 26, 418, 1992.

Fein, S. B., Lin, C. T., and Levy, A. S., Foodborne illness: perceptions, experiences, and prevention behaviors in the United States, *Journal of Food Protection,* 58, 1405, 1995.

Flynn, J., Slovic, P., and Mertz, C. K., Gender, race, and perception of environmental health risks, *Risk Analysis,* 14, 1101, 1994.

Frewer, L. J., Shepherd, R., and Sparks, P., The interrelationship between knowledge, control, and risk associated with a range of food-related hazards targeted at the individual, other people, and society, *Journal of Food Safety,* 14, 19, 1994.

Frewer, L. J. and Shepherd, R., Attributing information to different sources: effect on the perceived qualities of information, on the perceived relevance of information, and on attitude formation, *Public Understanding of Science,* 3, 385, 1994.

Grobe, D. and Douthitt, R., Consumer acceptance of recombinant bovine growth hormone: interplay between beliefs and perceived risks, *Journal of Consumer Affairs,* 29, 128, 1995.

Grobe, D., Douthitt, R., and Zepeda, L., Consumer risk perception profiles regarding recombinant bovine growth hormone (rbGH), *Journal of Consumer Affairs,* 33, 254, 1999.

Herrmann, R. O., Sterngold, A., and Warland, R. H., Comparing alternative question forms for assessing consumer concerns, *Journal of Consumer Affairs,* 32, 13, 1998.

Herrmann, R. O., Warland, R. H., and Sterngold, A., Nutrition concerns and food safety concerns occur independently among adults, *Journal of the American Dietetic Association,* 100, 947, 2000.

Hoban, T., Woodrum, E., and Czaja, R., Public opposition to genetic engineering, *Rural Sociology,* 57, 476, 1992.

Jordan, J. L. and Elnagheeb, A. H., Public perceptions of food safety, *Journal of Food Distribution Research,* 22, 13, 1991.

Jussaume Jr., R.A. and Higgins, L., Attitudes towards food safety and the environment: a comparison of consumers in Japan and the U.S., *Rural Sociology,* 63, 394, 1998.

Lin, C. J., Demographic and socioeconomic influences on the importance of food safety in food shopping, *Agricultural and Resource Economic Review,* 24, 190, 1995.

Mason, R., Boersma, L., and Faulkenberry, G. D., The use of open and closed questions to identify holders of crystallized attitudes: the case of adoption of erosion-control practices among farmers, *Rural Sociology,* 53, 96, 1988.

McGuirk, A. M., Preston, W. P., and McCormick, A., Toward the development of marketing strategies for food safety attributes, *Agribusiness,* 6, 297, 1990.

Nayga Jr., R. M., Sociodemographic influences on consumer concern for food safety: the case of irradiation, antibiotics, hormones, and pesticides, *Review of Agricultural Economics,* 18, 467, 1996.

Peterson, R. A., A meta-analysis of Cronbach's coefficient alpha, *Journal of Consumer Research,* 21, 381, 1994.

Sapp, S. G., Harrod, W. J., and Zhao, L., Social construction of consumer risk assessments, *Journal of Consumer Studies and Home Economics,* 18, 97, 1994.

Schafer, R. B., Schafer, E., Bultena, G. L., and Hoiberg, E. O., Food safety: an application of the health belief model, *Journal of Nutrition Education,* 25, 17, 1993.

Short, Jr., J. F., The social fabric at risk: toward the social transformation of risk analysis, *American Sociological Review,* 49, 711, 1984.

Sparks, P. and Shepherd, R., Public perceptions of the potential associated with food production and food production and food consumption: an empirical study, *Risk Analysis,* 14, 799, 1994.

Sterngold, A., Warland, R. H., and Herrmann, R. O., Do surveys overstate public concerns?, *Public Opinion Quarterly,* 58, 255, 1994.

Warland, R. H. and Herrmann, R. O., Do awareness filters identify knowledgeable respondents?, unpublished paper, 1999.

Chapter 10

Lessons Learned, Current Trends, and Future Needs

Elsa A. Murano and Neal H. Hooker

Multidisciplinary approaches to solving food safety problems draw from a diverse group of experts, involve people at various stages of the farm-to-table continuum and include industry, government, academia, and consumers. These three attributes are significant in that they explain what multidisciplinary approaches can do to enhance food safety. In this summary, will be discussed the above topics as they relate to the information contained in the preceding chapters. In addition, a discussion on the trends currently impacting the need for such multidisciplinary food safety research will be presented, including new trade barriers caused by food safety regulations, increases in litigation regarding product liability and plant closings, and the emergence of verification-based inspection systems. Finally, suggestions are provided for further developments of such multidisciplinary approaches as mechanisms that will facilitate formation of teams, as well as the resources necessary to support them.

0-8493-2217-0/01/$0.00+$.50

Lessons Learned

One of the principal lessons of the preceding chapters is that an interdisciplinary approach has the inherent capacity to succeed at solving problems in food safety where other types of approaches have failed. This is because: (1) such an approach draws from a diverse group of disciplines, improving the chances for finding successful solutions by looking at problems from more than one point of view; (2) the interdisciplinary approach involves people at various stages of the farm-to-table continuum, broadening the scope of solutions and applying them to all segments of the continuum in a way that does not antagonize or benefit one group at the expense of another; and (3) it includes industry, government, academia, and consumers, resulting in the unification of ideas through consensus building, and the commitment to support these through ownership by the various groups represented.

To illustrate the first point, the authors of Chapter 1 discussed the importance of including people from several disciplines in a Probabilistic Risk Assessment team. It is intuitively evident that calculating the risk of contamination of beef carcasses at a slaughterhouse with microbial pathogens necessitates the expertise of a mathematical modeling expert to build the model and a statistician to perform simulations with the model. However, microbiologists, epidemiologists, and veterinarians represent the disciplines that provide the data necessary to build the model in the first place, and the expertise to validate it once it is developed. In addition, having an economist on the team allows processors to determine how various levels of risk will affect costs, thereby serving as an incentive for the adoption of food safety systems such as HACCP.

Similarly, Chapter 2 depicted how the teaming of meat scientists and microbiologists provided useful comparison data regarding the effectiveness of several intervention strategies for carcass decontamination. More importantly, the inclusion of economists on the team added an extra dimension, providing data on the cost versus benefit of introducing each intervention into a slaughter plant. This information should not only help determine the best methods of decontamination, but also help specific processors select the type of technology that will improve the safety of their product in a way that is cost effective for their particular operation.

Illustration of the fact that expertise in multiple disciplines is essential for developing successful solutions to food safety problems is presented in Chapters 5 and 8. In Chapter 5, the authors show that only with a team composed of agricultural economists, food microbiologists, and epidemiologists can the situation of food safety in developing countries such as Brazil be accurately assessed. Further, the fact that very few such studies have been carried out underscores why food safety is such an elusive goal for such nations. Similarly, the authors of Chapter 8 note the need for microbiologists, social scientists, economists, marketing experts, and risk assessment experts to carry out meaningful studies on the impact of HACCP on specific sectors of the industry.

Even when studies are conducted where only a single aspect of a specific segment of the population is being considered, many experts are needed. We see an example of this issue in Chapter 9, where assessing the attitudes and opinions of consumers regarding food safety required a panel of economists, sociologists, psychologists, political scientists, food scientists, and marketing experts. The input of these individuals was necessary in order to design a questionnaire that would result in meaningful insights into the basis for food safety concerns expressed by consumers in the survey.

The second characteristic of an interdisciplinary approach is vertical integration across various segments of the farm-to-table continuum. Chapter 3 illustrates the importance of this by examining how the meat industry in various countries has formed vertical alliances between producers, processors, and retailers. In the U.K., for instance, the Tracesafe Cattle Management System traces the history of individual meat cuts to the animal of origin, making it easier to detect points of failure along the chain where contamination has occurred, and where systems can be introduced in the future to prevent this. Everyone benefits because producers, processors, and retailers depend on each other for the success of the program, and this is driven by their collective desire to reach lucrative export markets.

The third characteristic of an interdisciplinary approach is the involvement of various segments of the population, or stakeholders, in the discussion and implementation of solutions to the food safety problem. Chapter 7 illustrates this principle by showcasing the fact that, in the case of new technologies such as irradiation, the opinion of consumers must be well integrated with any laboratory studies of the method's effectiveness. Government officials, who make decisions regarding application of irradiation, need to be included as well, as their opinions are trusted more than those of industry by a significant portion of the consuming public. In addition, marketing experts are needed to work with consumers to identify the specifics of what these individuals are willing to accept, and what they are willing to pay for improved food safety.

Current Trends

We have seen from this summary evidence of the need for interdisciplinary approaches to addressing issues related to enhancing food safety. We should realize that several trends make it imperative to implement such approaches right now. One is the fact that food safety issues are significantly and increasingly affecting the way we do business with other countries. In Chapter 4, we read about phytosanitary barriers to international trade. Certainly, lack of compliance with requirements for a safe commodity prior to export can negatively affect the ability of any country to trade with others. Without an interdisciplinary approach, such an issue cannot be easily addressed due to the inherent complexity of the problem from both technical and legal viewpoints. Experts in various fields are needed (i.e., international law, microbiology, epidemiology, veterinary medicine, economics, finance, and political science) in order to assess the validity of disputes brought before international bodies such as the World Trade Organization, to make recommendations to policy makers on the fairest course of action based on sound science, and to give advice to governments and industry on the adoption of systems that will help them comply with international food safety requirements in a realistic, cost-effective, and successful manner.

Another trend contributing to the need for interdisciplinary teams is that of increasing litigation between consumers and industry, and between industry and regulatory agencies. Chapter 6 shows that lawsuits based on product liability issues related to foodborne illness outbreaks are on the rise. This is an issue that will not disappear as long as new threats to the safety of our food supply continue to emerge. The ability of microorganisms to adapt to food processing treatments designed to destroy them has resulted in outbreaks where conventional wisdom has failed to accurately predict the risk. On the other side of the coin, food processing companies have been closed down or levied fines by regulatory agencies when the blame may more correctly be placed on, or at least shared with, their suppliers. Such occurrences, especially in the U.S., have driven consumers to sue food processors, food companies to sue each other, and companies to sue regulatory agencies. Certainly, law suits have an effect on the economy of a nation, driving insurance costs up and market share and profits down. Interdisciplinary teams are needed to ensure that stakeholders understand the scientific basis of the problem of food contamination, and the responsibility that each segment of the farm-to-table continuum (including consumers, industry, and regulatory agencies) has to prevent foodborne illness.

A third trend that calls for the use of interdisciplinary teams is that of HACCP-based inspection, or verification-based inspection. Increasingly, governments are switching from a command-and-control system of inspection to one in which inspectors simply verify that plant personnel carry out their food safety preventive programs. Such a trend decries the need for interdisciplinary teams to be formed between industry, academia, and government agencies in order to provide producers and processors with much-needed information. Experts in microbiology, food processing, and statistical sampling and modeling are needed so that meaningful risk assessments are carried out as well as accurate monitoring and verification activities. In addition, marketing and economic expertise is needed so the true costs of implementing risk management systems such as HACCP can be ascertained ahead of time, and so product manufactured under methodologies that enhance safety are appropriately communicated and marketed to consumers. Further, effective risk communication throughout the farm-to-table continuum is necessary to ensure that all participants respond to their individual and group challenges. Currently, many processors lack some or all of this information, making the transition to a verification-based inspection system difficult. It is vital that further interdisciplinary food safety research is conducted to ease this transition.

Future Needs

Food safety is an issue that is here to stay. Throughout this book we have seen examples of how interdisciplinary teams have been used to address specific problems in this area. The next step should be to develop mechanisms that facilitate the formation of such teams with a minimum of bureaucracy. Certainly, there is a need for governments to step in and develop such teams for themselves, so that risk assessment can be conducted in an holistic and pragmatic manner. Thereby, the true impact of certain production and processing practices on the risk to the health of consumers can be determined and compared across risk factors, sources, and management strategies. Similarly, the food industry needs to develop mechanisms whereby teams that can address issues related to production and processing, from hazard analyses to economic analyses and crisis management, can be easily and quickly assembled. International trade organizations can also benefit from the adoption of such interdisciplinary approaches to settle disputes. To facilitate the process, alliances could be formed among various groups so that the building of interdisciplinary teams can be achieved with the kind of timing that is needed to meet the fast-paced changes in food safety.

It should be noted that in the academic realm, interdisciplinary teams should be the preferred way that scientists approach research problems in food safety. The lack of efficiency and breadth of expertise inherent in most conventional research should be replaced by forming teams of scientists with varied technological backgrounds. Along these lines, university researchers should lead the way by including both basic and applied scientists as well as extension specialists and consumer scientists in their interdisciplinary food safety research teams. Thus, their quest for knowledge can be enhanced by the combination of raw fact-finding with practicality, and by the rapid sharing of information with specific stakeholders.

It is important to realize that formation of such interdisciplinary teams, whether in industry, government, or academia, needs to be supported, not only in principle but also practically, through funds designated for such purposes. A serious effort must be made to designate such funding a high priority and to build incentives into funding evaluation reviews so that interdisciplinary approaches are favored over the conventional unilateral methods.

Finally, there is a need to be proactive in addressing future challenges in food safety. Interdisciplinary teams cannot be formed for the purpose of solving the here-and-now at the expense of planning for the future. Thus, these teams must devote significant efforts towards anticipating the trends that will impact food safety in the next ten years. One way to do so is to build in linkages among governments for the purpose of sharing information regarding outbreaks and the risks to food safety posed by emerging pathogens in various regions of the world and in specific commodities. Another strategy is to include trend analysts and forecasters in interdisciplinary teams, so that accurate predictions can be made and followed up through appropriate resource allocations.

It is clear from the information provided in this book that interdisciplinary approaches are the best approaches to addressing problems in food safety. Evidence has shown this to be the case. However, working with individuals from various disciplines in a team requires effort, coordination, and good planning. Without such interdisciplinary food safety research teams, many pressing concerns cannot be addressed. Therefore, the question we should ask is whether we have the commitment necessary to follow through with such an approach. If we do, we know that the benefits will outweigh any inconveniences. More importantly, we know that to fail in developing interdisciplinary teams will mean to fail in our quest for enhancing the safety of our food supply.

Index